DISCARD

North Providence Union Free Library
1810 Mineral Spring Avenue
North Providence, RI 02904
(401) 353-5600

The Extreme Earth

Ocean Ridges and Trenches

Peter Aleshire

Foreword by
Geoffrey H. Nash, Geologist

OCEAN RIDGES AND TRENCHES

Copyright © 2007 by Peter Aleshire

All rights reserved. No part of this book may be reproduced or utilized in any form or by any means, electronic or mechanical, including photocopying, recording, or by any information storage or retrieval systems, without permission in writing from the publisher. For information contact:

Chelsea House
An imprint of Infobase Publishing
132 West 31st Street
New York NY 10001

ISBN-10: 0-8160-5919-5
ISBN-13: 978-0-8160-5919-5

Library of Congress Cataloging-in-Publication Data
Aleshire, Peter.
　Ocean ridges and trenches / Peter Aleshire; foreword, Geoffrey H. Nash, geologist.
　　p. cm. — (The extreme earth)
　Includes bibliographical references.
　ISBN 0-8160-5919-5
　1. Ocean bottom—Juvenile literature. 2. Mid-ocean ridges—Juvenile literature.
　3. Deep-sea ecology—Juvenile literature. I. Title.
　GC87.A44 2007
　551.46'8—dc22   2006032058

Chelsea House books are available at special discounts when purchased in bulk quantities for businesses, associations, institutions, or sales promotions. Please call our Special Sales Department in New York at (212) 967-8800 or (800) 322-8755.

You can find Chelsea House on the World Wide Web at http://www.chelseahouse.com

Text design by Erika K. Arroyo
Cover design by Dorothy M. Preston/Salvatore Luongo
Illustrations by Richard Garratt

Printed in the United States of America

VB FOF 10 9 8 7 6 5 4 3 2 1

This book is printed on acid-free paper.

Contents

Foreword by Geoffrey H. Nash vii
Preface ix
Introduction xi

Origin of the Landform: Ocean Ridges and Trenches 1

1 ⋄ Mid-Atlantic Ridge, Atlantic Ocean 5
   Alexander the Great: The First Deep-Sea Diver 6
  Jigsaw Puzzle World 6
  An Ocean of Mystery 8
  Refining a Crazy Theory 9
   The Descent of the Bathysphere 12
  Seafloor Mysteries Mount 13
   Mid-Atlantic Ridge 14
  Dusting Off an Old Theory 14
   Where Giant Squid Lurk 15
   Puerto Rico Trench: Deepest Atlantic Ocean Trench 17
  Mid-Atlantic Ridge Mapped 17

2 ⋄ San Juan de Fuca Ridge, Pacific Ocean 20
  Probing the Magnetic Field 22
  Mysterious Magnetic Stripes 23
   Earth's Poles Flipping? 25
  Paper Models and Cracks in the Earth 26
   The San Andreas Fault 27
  Unraveling the Mystery 29
   A Woman Joins the Boys' Club of Science 29
   Mount St. Helens 32
  Plate Tectonics Triumphs 33

3 ✧ The Mariana Trench, Pacific Ocean — 35
- *Pacific Ocean: Vital Statistics* — 37
- History of a Dream — 37
  - *Auguste Piccard (1884–1962)* — 38
  - *The Pacific Basin* — 39
- Into the Deep — 42
  - *The Mystery of Seamounts* — 44
- Terrible Explosion Resounds — 45

4 ✧ The Galápagos Rift, Pacific Ocean — 48
- Surprise on the Seafloor — 50
  - *The Theory of Evolution* — 52
- Bewildering Creatures Discovered — 53
  - *Strange Reproduction in the Dandelion Patch* — 54
- A World Run on Sulfur — 55
  - *Life Runs on Sunlight* — 56
  - *Did Life Originate along the Vents?* — 58

5 ✧ East Pacific Rise, Pacific Ocean — 61
- *Measuring Gravity* — 62
- Race to Explore a Bizarre World — 64
  - *The Making of Gold and Silver* — 66
- Mounting a Historic Expedition — 66
  - *Lakes of Lava* — 69
- Nasty Surprise Awaits — 69
  - *Where Does the Ocean Get Its Salt?* — 70

6 ✧ Arctic Ridge, Arctic Ocean — 73
- *The Arctic Ocean* — 73
- *A Wandering Pole* — 76
- Daring Explorer Braves the Ice — 76
  - *Explorer and Humanitarian* — 77
- Arctic Survey Produces Surprises — 78
- Volcanoes Challenge Theories — 80
  - *How Undersea Ridges Can Affect the Climate of a Planet* — 82
- Odd Rocks Lubricate Fissures — 83

7 ✧ Iceland, North Atlantic Ocean — 84
- *The First Icelanders* — 85
- The Fury of the Earth Spirits — 85

A Hot Spot or Something Else? 87
   The Deccan Traps 87
   The Hawaii Hot Spot 89
What Drives Hot Spots? 90
Icelanders Live on Fire's Edge 91
   Strange Life Abounds 92
Volcano Threatens Town 93

8 ✧ The Java Trench, Indian Ocean 95
   Java Trench (Sunda Double Trench)
   Vital Statistics 95
Girl's Geography Lesson Saves Lives 96
   The World's Worst Tsunamis 97
   Disaster Warning System 98
Wave Extracts Terrible Toll 99
Trenches Mirror Ridges 100
Buried Plate Melts 101
History's Worst Volcano 104
   Krakatoa: When Lava Flies 107
Ash Affects Climate 108
   The World's Worst Earthquakes 108

9 ✧ Peru-Chile Trench, Pacific Ocean 110
   The Tsunami of 1868 111
Collision of Plates Creates Andes 112
Climate Can Affect Mountain Building 114
   Stuck in a Trench 115
Trenches Accumulate Nutrients 115
   Search for the Origins of the Andes 117
   The Nitrate Mystery 119

10 ✧ Red Sea, North Africa and the Middle East 120
Red Sea Nourished Civilization 122
Creating a New Trench? 125

Glossary 128
Books 134
Web Sites 137
Index 139

Foreword

Mid-ocean ridges and trenches are the biggest of Earth's geologic landforms that you will never see. They are completely hidden from view, deep beneath the surface of the ocean. Scientists have visited them in submersibles and studied them via sonar. These large-scale features are the product of the hot, semi-molten crustal material deep within the Earth. The mid-ocean ridges are constantly growing, and the trenches are constantly recycling the older ocean-crust rocks. Sitting on top of the ocean crust and poking above the surface of the watery oceans are the lighter rocks of the continents on which we all live. The mid-ocean ridges and trenches of the Earth are the telltale signs of the theory of plate tectonics that was only first described in the early 1960s. In spite of the relatively short period of time that the theory has been around, it has made a monumental shift in our understanding of how the slow processes of the ever-changing Earth work. The test of a scientific theory is whether it explains what is found in the field or laboratory, and with plate tectonics, it all dropped into place.

These geologic features are directly responsible for some of the most frightening events we can witness: volcanoes and earthquakes. They are also the force behind all of the mountain-building events that have occurred during the 4.5 billion years of the Earth's history.

Ocean Ridges and Trenches by Peter Aleshire presents examples of ridges and trenches shaped by the powerful forces of plate tectonic activity. This book visits 10 unforgettable locales around the world, describing their power and global processes at work beneath our feet. You may be familiar with some of the places mentioned, such as the Mariana Trench, the deepest point in the ocean, or the San Andreas Fault in California, the fault responsible for so many of California's earthquakes. You can also find answers to such questions as where the oceans get their salt.

Studying these exceptional examples can provide an understanding of how mid-ocean ridges and troughs develop and change over the span of geologic time. This book also conveys the scope of geologic time in which these natural processes occur and how so many scientists have contributed to our current understanding. Without the theory of plate tectonics and the knowledge of mid-ocean ridges and troughs, scientists would be trying to fit all of their current observations and data into a static-state model of the Earth. With the understanding of the mechanisms of ridges and troughs, the surface of the Earth as it looks today can be explained and predictions about its future contours can be made.

Two recurring features of this book are the author's focus on the inquisitive scientists who have spent their careers researching and also the author's insights into the consequences for humans that result from volcanoes, earthquakes, and tsunamis. Scientists who study these mid-ocean ridges and trenches are geologists, biologists, seismologists, and engineers. They study the molten rocks that form them, the life that makes them their homes, and even the cultures of people that live in close proximity to them.

Ocean Ridges and Trenches should be your reference into the natural processes of ridges and trenches and a window into the driving force behind volcanism, mountain building, and earthquakes.

—Geoffrey H. Nash, geologist

Preface

From outer space, Earth resembles a fragile blue marble, as revealed in the famous photograph taken by the *Apollo 17* astronauts in December 1972. Eugene Cernan, Ronald Evans, and Jack Schmitt were some 28,000 miles (45,061 km) away when one of them snapped the famous picture that provided the first clear image of the planet from space.

Zoom in closer and the view is quite different. Far beneath the vast seas that give the blue marble its rich hue are soaring mountains and deep ridges. On land, more mountains and canyons come into view, rugged terrain initiated by movement beneath the Earth's crust and then sculpted by wind and water. Arid deserts and hollow caves are here too, existing in counterpoint to coursing rivers, sprawling lakes, and plummeting waterfalls.

The Extreme Earth is a set of eight books that presents the geology of these landforms, with clear explanations of their origins, histories, and structures. Similarities exist, of course, among the many mountains of the world, just as they exist among individual rivers, caves, deserts, canyons, waterfalls, lakes, ocean ridges, and trenches. Some qualify as the biggest, highest, deepest, longest, widest, oldest, or most unusual, and these are the examples singled out in this set. Each book introduces 10 superlative examples, one by one, of the individual landforms, and reveals why these landforms are never static, but always changing. Some of them are internationally known, located in populated areas. Others are in more remote locations and known primarily to people in the region. All of them are worthy of inclusion.

To some people, the ever-shifting contours of the Earth are just so much scenery. Others sit and ponder ocean ridges and undersea trenches, imagining mysteries that they can neither interact with nor examine in person. Some gaze at majestic canyons, rushing waterfalls, or placid lakes, appreciating the scenery from behind a railing, on a path, or aboard a boat. Still others climb mountains, float rivers, explore caves, and cross deserts, interacting directly with nature in a personal way.

Even people with a heightened interest in the scenic wonders of the world do not always understand the complexity of these landforms. The eight books in the Extreme Earth set provide basic information on how individual landforms came to exist and their place in the history of the planet. Here, too, is information on what makes each one unusual, what roles they play in the world today, and, in some cases, who discovered and named them. Each chapter in each volume also includes material on environmental challenges and reports on science in action. All the books include photographs in color and black-and-white, line drawings, a glossary of scientific terms related to the text, and a listing of resources for more information.

When students who have read the eight books in the Extreme Earth set venture outdoors—whether close to home, on a family vacation, or to distant shores—they will know what they are looking at, how it got there, and what likely will happen next. They will know the stories of how lakes form, how wind and weather work together to etch mountain ranges, and how water carves canyons. These all are thrilling stories—stories that inhabitants of this planet have a responsibility to know.

The primary goal of the Extreme Earth set of books is to inform readers of all ages about the most interesting mountains, rivers, caves, deserts, canyons, waterfalls, lakes, ocean ridges, and trenches in the world. Even as these books serve to increase both understanding of the history of the planet and appreciation for all its landforms, ideally they also will encourage a sense of responsible stewardship for this magnificent blue marble.

Introduction

A frightened 10-year-old watches as the ocean retreats far from the beach and, remembering her geography lesson, saves her family from an approaching tsunami. A German weatherman puzzles over reports of strange matching fossils in North America and Europe and comes up with an idea that revolutionizes our understanding of the Earth. A determined explorer in a damaged submarine lands in the world's deepest place and immediately confronts the astonished gaze of a fish that should not be there. A sea-loving scientist with a yen for adventure journeys to the ocean bottom seeking answers to interesting questions but uncovers a stunning mystery that may reveal the origins of life. A tough, obsessively curious man maroons his wood-hulled ship in the ice at the top of the world to solve mysteries that end up altering our ideas as to how the Earth works.

These are all people who played a vital role in one of the great scientific revolutions in human history, which involved the world miles beneath the ocean surface at the base of the planet's most awe-inspiring chain of mountains and canyons.

A vast chain of mountains snakes down the middle or along the edge of every one of the world's oceans. Some of those mountain chains rival the highest mountains on the continents in height, and they all dwarf the terrestrial mountain chains in length. Those mountain chains are echoed by unimaginable canyons in the seafloor, some plunging to seven miles (11.3 km) deep and running in the blackness of the deep sea for thousands of miles. This network of mid-ocean ridges and undersea trenches marks the edges of nearly two-dozen great crustal plates that have controlled the evolution of the planet as well as of human beings.

Ocean Ridges and Trenches will focus on the 10 most unusual ridges and trenches and the remarkable people who struggled to uncover the relationship between them as well as their role in the world of science. Driven by an insatiable need to understand, scientists come up with theories that explain the bewildering behavior of the world. They then go out to find the facts necessary to prove their theories. That intimate, complicated, vital connection between a good theory and better evidence drives science.

The attempt to solve the riddle of the ridges led directly to the theory of plate tectonics, which in turn helped explain tsunamis, earthquakes, volcanoes, the Earth's climate, and the evolution of life on this planet. The titanic forces that created the ridges and trenches have inflicted terrible devastation, such as the tsunami in 2004 that killed more than 150,000 people on islands and continents bordering the Indian Ocean. But those same forces have produced all the precious metals and rare elements on which our economies and living things depend. They also have controlled the climate, the evolution of life, and the rise of civilizations.

So the discovery of underwater chains of mountains 50,000 miles (80,500 km) long and gashes seven miles deep on the seafloor represents one of the great triumphs of science in human history. And the role of the remarkable scientists and explorers who solve the mystery makes a great adventure yarn and a testament to human daring and yearning.

Origin of the Landform

Ocean Ridges and Trenches

The planet's most dramatic, massive, and revealing geological features are almost all hidden from view on the seafloor, usually miles beneath the sunlit surface. Undersea ridges include mountains taller than Mount Everest in a nearly continuous chain some 50,000 miles (80,470 km) long. Those chains of underwater volcanoes are echoed by a system of trenches or canyons—some seven miles (11.3 km) deep. Almost all of these underwater features are marked by volcanoes, earthquakes, and fresh, volcanic basalt that are generally much younger than most of the rocks on the continents.

The origin of this remarkable system of ridges and trenches lies deep inside the Earth and is intimately connected to almost every feature of the surface of the planet, from the existence of the continents to the retention of a breathable atmosphere. These long chains of mountains and deep, narrow canyons are caused by the basic physics of the Earth's structure. Scientists studying the change in the speed of energy waves generated by earthquakes as they pass through the layers of the Earth have gained a general picture of the structure of the planet, even thousands of miles beneath the surface.

The story starts in the Earth's solid, mostly iron inner core, a sphere about the size of the Moon. The inner core is about 1,500 miles (2,400 km) in diameter and rotates at a slightly different rate than the surrounding planet. It has been heated to an estimated 7,772°F (4,300°C), mostly by the natural radioactive decay of elements in rocks. The core would be boiling molten rock if it were not for the enormous pressure of more than 3,000 miles (4,800 km) of overlying rock. But the heat of the dense inner core radiates into the outer core, which can turn to liquid despite a lower temperature because of the reduced pressure imposed by the thinner layer of overlying rock.

The molten outer core is 4,200 miles (6,800 km) in diameter and composed mostly of iron and sulfur, heated to a temperature of roughly

6,700°F (3,700°C). Because the outer core is only 1,800 miles (2,900 km) beneath the surface and therefore under less pressure, the molten rock can boil and flow in great convection currents. Currents in this area of the outer core probably generate the Earth's magnetic field. These circular roils of molten rock transmit energy and movement to the Earth's next layer, the semi-molten mantle, which contains most of the Earth's mass.

The roughly 1,800-mile- (2,900-km-) thick mantle is made of lighter rocks than the iron-rich core, including aluminum, magnesium, oxygen, silicon. Here massive, slow-motion convection currents transfer heat from the bottom of the mantle toward the Earth's surface. The rocks of the mantle ooze and flow at temperatures of 1,800–3,600°F (980–1,980°C), which means the current flows along at maybe an inch (2.5 cm) per year. Geophysicists estimate that these convection cells in the mantle may have only completed four to six rotations in the past 4 billion years of the Earth's history.

Nonetheless, the inexorable movement transfers energy and pressure to the thin, brittle, outermost layer of the Earth, the crust. The crust is about 3.7 miles (6 km) thick beneath much of the ocean floor and about 37 miles (60 km) thick beneath the continents. The rocks of the crust contain the continents and ocean basins, making it possible for life to survive on the surface. But the crust must constantly absorb the energy from those slow-motion currents in the underlying mantle.

This results in the development of undersea ridges and trenches, not to mention the configuration of the continents. The upwelling of semi-molten magma in the mantle creates a great crack in the crust along the rising wall of the convection current. Magma wells into the crack from the upper mantle to push apart the crust and create a continuously roiling chain of volcanic activity that creates the long chain of undersea ridges.

On the other side of this current in the mantle, the now cooler, sinking wall of the convection cell drags the brittle crust with it. The fissure along which this captured piece of crust descends into the mantle creates the seam of an undersea trench.

So the magma welling up from the mantle that creates new crust on the floor of the ocean along an undersea ridge is pushed outward from the ridge and across the ocean basin until it finally encounters an undersea trench, where it is forced back down toward the mantle where it is remelted and recycled. This system of cracked crustal rock moving between oceanic ridges and trenches divides the entire surface of the planet into huge, splintered chunks of rock called crustal plates.

That accounts for the undersea ridges and trenches, but it does not account for the continents, which are made of the lightest rocks of all. The continents effectively float atop the dense rock of the seafloor.

Usually, the rocks of the continents are too light and buoyant to get drawn down into the trenches, so they can move about the surface embedded in the dense rock of the ocean crust.

The forces that created ridges and trenches can be followed down to the very core of the Earth, a great boil of molten rock that makes life on the cool surface of the planet possible.

1

Mid-Atlantic Ridge
Atlantic Ocean

In 1911, German meteorologist Alfred Lothar Wegener sat quietly in the silence of a great library, pondering a solution to a vexing mystery. Why did a set of fossils in North America so perfectly match the fossils in Europe? Of course, he was a weatherman, not a geologist or a paleontologist, so some said he had no business even asking the question, much less suggesting a theory about how the Earth fit together that would explain the oceans, the continents, earthquakes, *volcano*es, misplaced fossils, mismatched mountain ranges, and the structure of the Earth.

In fact, Wegener was something of an adventurer and a scientific dabbler. He received his Ph.D. in astronomy from the University of Berlin in 1904. But then he got interested in geophysics and began focusing on the climate. Wegener came up with the brilliant idea of using hot-air balloons to trace wind currents in the upper atmosphere. Trudging across the vast expanse of ice in Greenland, he developed theories on how climatic changes at the top of the world generated weather all over the planet and then wrote a brilliant textbook on the weather. He was a bright, creative, adventuresome man who did not stick to his intellectual cubbyhole.

Still, Wegener did not know much about fossils and had no good reason to be reading the scientific paper on fossils he stumbled across in that library. But he could not help but notice something strange about the comprehensive list of fossils of creatures that lived when dinosaurs roamed the Earth. It looked like almost identical creatures lived in Europe and North America some 300 million years ago. That seemed odd. So he looked further and saw a baffling matchup between the fossils in Africa and South America. This also seemed strange. Then he found something even more peculiar: Someone had found fossils of tropical plants on the arctic island of Spitsbergen. How could tropical plants possibly endure the cold? Could the climate have changed so much in 300 million years?

The more he studied, the more puzzles he uncovered. Rocks on opposite sides of the Atlantic Ocean seemed to mirror one another. For

> ## ALEXANDER THE GREAT: THE FIRST DEEP-SEA DIVER
>
> Many historians believe that Alexander the Great, the insatiable conqueror of much of the known world, was the first deep-sea diver. Reportedly, the youthful conqueror ordered his wizards and alchemists to construct a glass barrel sometime in the fourth century B.C.E. He then climbed into the glass barrel and commanded his advisers to lower him beneath the surface. Dangling far beneath the surface, the warrior who would one day conquer much of the known world waited patiently to see what monsters would swim past. Upon his return, he supposedly described a fish so big that it took three days to swim past, fueling the ancient fascination with sea monsters. Of course, either he glimpsed the last example of some unimaginably large creature or he indulged in a human tendency to exaggerate his exploits. Today biologists believe that the 100-foot- (30-m-) long blue whale is the largest creature that has ever lived.

instance, rocks in a portion of the Appalachian Mountains in the United States precisely matched the age and composition of rocks in the Scottish Highlands. Meanwhile, a distinctive layer of rocks in South Africa also perfectly echoed layers in Brazil.

JIGSAW PUZZLE WORLD

While studying a map of the world, Wegener thought it odd that the northern projection of Africa fit so neatly into the southern swoop of South Africa. Moreover, the coast of Europe and England seemed to match up with the coast of North America. Of course, people had noticed the tantalizing fit of the continents going back to the 16th century. And a few years earlier, Austrian geologist Eduard Suess had suggested that a single great continent he dubbed Gondwanaland had covered most of the planet before cracking apart. Suess hypothesized that some sections sank to form the great basin of the Atlantic Ocean. He maintained that the Earth had gradually cooled, cracked, and contracted, wrinkling like the surface of a dried-up apple and creating the great mountain chains and *ocean basins* in the process.

But now Wegener had a strange idea. Suppose the continents had once huddled together in some kind of *supercontinent*. Then suppose that supercontinent split apart and the pieces went drifting across the seafloor to their present locations. That would explain his otherwise puzzling observations. The rocks matched because they were made at the same time in the same place before dispersing. The fossils matched because 300 million years ago Europe, North America, South America, Africa, Australia, Asia, and even arctic islands were all part of a single, giant continent on which the dinosaurs first arose somewhere near the equator. "A conviction of the fundamental soundness of the idea took root in my mind," Wegener later wrote.

Alfred Wegener's 1915 reconstruction of continental drift

The experts immediately dismissed his theory. Other people had noticed the corresponding fossils around the world, but they accepted the idea that *land bridges* must have once connected the continents, stretching across the oceans, just like the Bering Strait by Alaska. Maybe those hidden land bridges rose to become dry land during *ice age*s, when sea levels dropped hundreds of feet around the world because so much water froze into ice at the poles. This would allow animals to move from continent to continent along land bridges that had since sunk beneath the ocean, argued the fossil experts. After all, you could hide almost anything in the ocean.

AN OCEAN OF MYSTERY

In 1911, the oceans covering three-quarters of the Earth's surface remained an absolute mystery. The ocean covered 140 million square miles (363 million km^2), but no one knew how deep it was. A few explorers had dropped weights on long lines, demonstrating that it must be at least several miles deep, but no one had ever found the deepest point. So those drowned land bridges could be out there, the ocean's secret. The experts advised Wegener to not waste his time on his preposterous theory. They had all kinds of interesting ideas to account for the formation of the deep oceans and the high continents without suggesting anything so foolish as the continents wandering, pushed by some mysterious force.

Most European geologists accepted Suess's theory of the contracting, wrinkled Earth. That is what they believed created the great mountain ranges and the ocean basins.

Most American geologists accepted a different version of the theory, developed by geologist James Dwight Dana. He believed the continents formed first, since they were made of quick-cooling rocks such as quartz and feldspar. The ocean basins cooled more slowly, since they were composed of slower-cooling *olivine* and pyroxene. The different cooling rates caused different rates of contraction, accounting for the deep ocean basins and the high-riding continents. They could even explain the giant mountain ranges that run along the edge of many continents, such as the Andes and the Himalayas. Surely those gigantic mountain ranges had puckered up along the margins between the quick-cooling continents and the slow-cooling ocean basins.

Other geologists at the time argued that the ocean basins were left over from an exceptional event that took place as the molten Earth cooled. The spin of the Earth had perhaps set up waves in the still molten rock. The waves circled the globe, building up on top of each other. At some point, just as the Earth's surface solidified, this rotation-driven wave of semi-molten rock ripped loose a great chunk of the Earth, which spun off into space. That created the Moon and left behind a great hole, which became the *ocean basins*. In any case, hardly anyone besides Wegener believed that continents could go skittering across the seafloor.

REFINING A CRAZY THEORY

Wegener spent years refining his theory. He accumulated the lists of fossils and rocks. He matched up the edges of the continents along their continental margins, ledges of accumulated continental sediment just off the coasts, and found an even better fit between the continents.

He had still not finished work on his theory when World War I broke out. He was drafted into the German army, which hurled itself against the French and the British in a terrible bloodbath that shaped the 20th century. Wegener was badly wounded during one of the bloody battles on the western front. He lay for months in a hospital, slowly recovering. Lying in his hospital bed, his mind ran back and forth over the evidence that continents have wandered the surface of the planet for millions of years. When he recovered, he became a weatherman for the army.

After Germany's bitter loss in that global war, Wegener returned to the university as a professor. There he resumed work on the theory he would dub "continental drift." He first published the theory in 1915 and expanded on it throughout the 1920s.

He dubbed the supercontinent *Pangaea*, which in Greek means "all the Earth." He said Pangaea shattered and the pieces drifted off through the ocean *crust* to their present locations, moving at about 10 inches (25 cm) per year, like icebreakers plowing through the ice.

The experts mocked him. Dr. Rolling T. Chamberlin of the University of Chicago observed, "Wegener's hypothesis in general is of the footloose type, in that it takes considerable liberty with our globe and is less bound by restrictions or tied down by awkward, ugly facts than most of its rival theories."

Wegener's critics pointed out that he had offered no explanation for how the continents could wander. He had vaguely cited the spin of the Earth combined with the gravitational tug of the Moon, but that was implausible. Such a massive tidal force from the Moon would stop the Earth from spinning within about a year. Moreover, if the continents were plowing through the ocean floor like icebreakers, their edges would be so smashed up that they could not possibly still fit together as Wegener suggested.

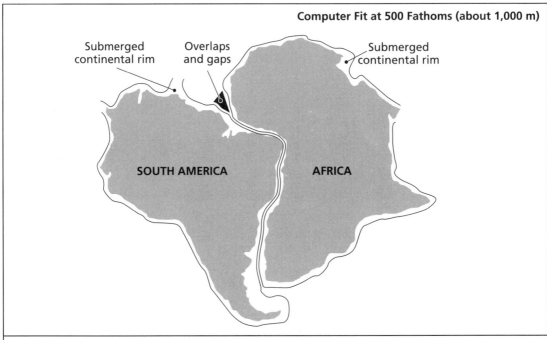
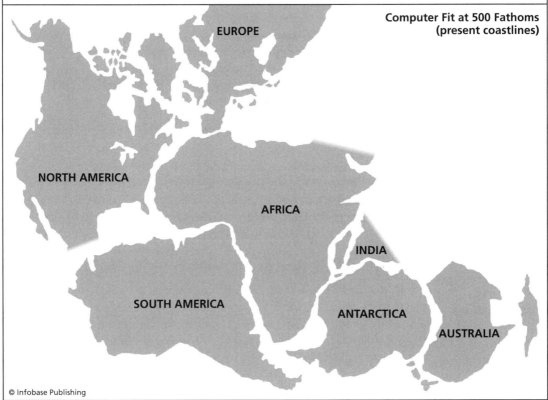

Evidence for continental drift

Granted, some geologists cautiously suggested that bumper-car continents might explain the remarkable, crunched-up rock layers of the Alps and the jigsaw puzzle pattern of the fossils, but most geologists dismissed the theory as a wild, physically impossible effort to turn the coincidental fit of the continents into a theory.

Spurned by his colleges, Wegener returned to his first obsession, understanding polar weather patterns. He still believed in his theory and in combining insights from different fields: "Scientists still do not appear to understand sufficiently that all Earth sciences must contribute evidence toward unveiling the state of our planet in earlier times, and that the truth of the matter can only be reached by combining all this evidence. It is only by combining the information furnished by all the Earth sciences that we can hope to find the picture that sets out all the known facts in the best arrangement and that therefore has the highest degree of probability. Further, we have to be prepared always for the possibility that each new discovery, no matter what science furnishes it, may modify the conclusions we draw."

He returned to Greenland in 1930 to study the weather there. When another group of scientists got stranded on the ice, Wegener led an expedition to bring them food. On his way back across the ice, he froze to death the day after his 50th birthday, a brave and creative scientist ridiculed and rejected by the experts.

Fortunately, that is not the end of his story; it is the start of the most dramatic, productive, and revolutionary era in the history of the Earth sciences. Wegener laid the foundation for astonishing discoveries that revealed the nature of the Earth, vast chains of undersea mountains, and deep trenches. In the end, he helped roll the first rock down a slope that set loose an avalanche of discovery and transformed humans' view of the planet.

The problems with the conventional explanation for mountains and the ocean basins began to slowly pile up, even as Wegener's theory sank into obscurity. For instance, geologists carefully measured the crunched and folded rock layers that constituted the Alps and the Appalachians. They discovered that hundreds of miles of rock layers had been folded sideways and compressed into the mountains. That seemed like far too much crunching and compressing to be explained by the mere cooling and contraction of the Earth.

Next, surveyors such as George Everest, struggling to measure the exact heights of the world's tallest mountains, uncovered another mystery. They discovered that the mass of rock in the mountains had enough gravity to affect measurements made by surveyors. But when they tried to compensate for the effect of the mountain's gravity, they discovered that massive mountains such as Mount Everest had only about half the

THE DESCENT OF THE BATHYSPHERE

The year Wegener died on the ice, two brave men diving in the warm waters of Bermuda launched the new era of deep-sea diving. Charles William Beebe and Otis Barton squeezed into the first diving bell and descended 1,426 feet (435 m) into the water, three times deeper than any previous diver.

Beebe was a poet, showman, and explorer who had tracked rare birds in the Tropics, hiked up an erupting volcano, and in the 1920s fashioned his own diving helmet. He had once weighed himself down and walked the ocean bottom close by shore. He wrote enthusiastically about his adventure. "Don't die," he wrote, "without having borrowed, stolen or made a helmet of sorts, to glimpse for yourselves this new world."

Reading Beebe's words, Otis Barton did as he advised. He made his own crude helmet out of a wooden box with glass windows, weighed himself down with sandbags, and explored the bottom of the harbor of Cotuit, Massachusetts, while a friend pumped air down to him with a bicycle pump.

He eventually sought out Beebe and convinced the older man that they could dive together to great depths in a steel sphere with two windows of fused quartz that Barton insisted would withstand the enormous pressure of the deep ocean.

They constructed a five-ton steel sphere with 1.5-inch- (3.8-cm-) thick walls that would dangle at the end of a 3,500-foot (1,070-m) cable. On the first test dive, the *Bathysphere* nearly became their coffin when it began spinning as it descended to 2,000 feet (610 m), almost snapping the support cable. Undaunted, they redesigned the cable system and descended again, the first human beings to explore the depths of a realm as mysterious and forbidding a distant planet.

As they descended through light into darkness, the ocean around them turned to a deep, cold, mystical blue. "The blueness of blue passed into our very beings," wrote Beebe. "It seemed to me that it must be like the last terrific upflare of a flame before it is quenched."

They brought back rapturous descriptions of the creatures they glimpsed, many of them glowing with their own *bioluminescence,* a glow-in-the-dark substance produced by microscopic organisms that sometimes illuminates breaking waves at night. Deep-sea creatures use these naturally glowing materials and bacteria to communicate in the absolute darkness. Barton and Beebe saw clouds of jellyfish, mysterious flashing patterns of light, and strange fish. At one point, a length of rubber hose worked loose and drifted past their window, backlit by invisible, glow-in-the-dark plankton. The enthralled explorers momentarily thought the bizarre shape was a huge, distant sea monster. They also glimpsed glowing hatchet fish, two-inch- (5-cm-) long creatures sporting gaping, needle-toothed jaws.

They stopped at 1,426 feet (435 m), still far above the bottom, fearful that the enormous 900 pounds (408 kg) per square inch of pressure would burst the seams of the bathysphere, which was already leaking from some unknown seam. Each of the quartz windows held back nine tons of pressure, equivalent to the weight of five cars on top of a window the size of a computer monitor.

Barton and Beebe continued to dive for several more years, eventually reaching a depth of 3,028 feet (923 m). Beebe published numerous articles and a vivid book, *Half Mile Down,* in 1934. He described many marvels, including a six-foot- (2-m-) long "sea dragon" and a large "constellationfish," with glowing rows of yellow dots and intensely purple lights, which has never been glimpsed since.

While many of Beebe's observations have remained controversial, he and Barton launched the modern era of deep-sea exploration, inspiring a coming generation of scientists who would discover in the supposedly featureless depths of the ocean a realm of surprise and mystery, not to mention the cradle of life itself.

mass they expected. Apparently, the mountains were made of light rock with deep roots, so the mountain was "floating" like an iceberg on top of dense rock. Again, that observation conflicted with the idea of a uniform, shrinking Earth.

Finally, geologists discovered that many rocks have natural radioactivity. This gradual decay of elements in the rocks generates heat. But that means the Earth is not cooling and shrinking fast enough to wrinkle up into mountain ranges and ocean basins.

Although most geologists still struggled to find a way to make the old theories work, some begin to reconsider Wegener's theory of continental drift. Some geologists suggested that perhaps the continents could be moved as a result of something happening in the hot, molten and semi-molten layers beneath the crust. Maybe the heat of the *radioactive* rocks could melt the crust, so continents could slide along. After all, a thick slab of glass can flow very slowly without becoming a liquid. Maybe the continents could do the same thing.

SEAFLOOR MYSTERIES MOUNT

In the meantime, other baffling clues began to emerge from the seafloor itself. Once, scientists had assumed that the deep oceans were flat, featureless, and lifeless, great plains buried beneath the mud flowing off the continents. They believed the ocean bottoms were too cold, deep, and lightless to sustain life. Early attempts to find the bottom with cannonballs on ropes and cables only demonstrated that across vast mid-ocean stretches, the oceans were miles deep. Some soundings, however, suggested that undersea mountains rose in some places, including an intriguing chain of underwater peaks in the middle of the Atlantic Ocean.

The first effort to explore the world's oceans on a scientific basis dates back to the epic, three-year voyage of the HMS *Challenger* in 1872. The *Challenger* circled the globe and tried to measure the depth of the ocean once every 100 miles (161 km) with a 200-pound (91-kg) weight attached to a hemp line connected to a hand-operated winch.

The first hints of a strange mountain range in the middle of the Atlantic Ocean came in the mid-1800s when Lieutenant Matthew Fontaine Maury set out to make as many measurements as possible and combine them with hundreds of soundings by navy vessels and fishermen. He produced the first, crude contour maps of the Atlantic Ocean that revealed a long, fitful, shadowy mountain range miles beneath the surface. He called it "Middle Ground" or "Dolphin Rise." However, he was frustrated by the great gaps in the knowledge of the seafloor.

Explorations of the ocean floor made a gigantic leap in the 1920s when the navy began to experiment with making crude depth maps by

MID-ATLANTIC RIDGE	
length:	42,000 miles (67,600 km), intermittent
area:	one-quarter of the Earth's surface
width:	600–2,500 miles (970–4,020 km)
Mid-Atlantic Ridge section:	12,000 miles (19,300 km) long
topography:	topped by a mile-deep rift wider than the Grand Canyon, with sheer walls 1,000 feet (305 m) high in places

bouncing pulses of sound off the seafloor, then waiting for the echoes to return to the ship. By calculating the travel time of the signal, scientists could finally get a rough measurement of the depth of the ocean across wide swaths.

Other scientists developed instruments that could directly measure the force of gravity. Once they figured out how to tow those instruments behind ships, they discovered strange decreases in gravity readings in the ocean.

DUSTING OFF AN OLD THEORY

The accumulation of odd measurements prompted some scientists to dust off theories of wandering continents to see if they could think of some physical force that would move the continents. University of California professor David Griggs made a model of the Earth using a tank full of oil to represent the deep, fluid layers of molten rock. He covered the oil with a thin film of wax, to represent the hard, solid crust of the Earth. Then he used rotating barrels to create simulated *convection currents*, similar to the roiling boil of a pan of water. Sure enough, the slow currents in the oil moved the paraffin layers. Perhaps currents in the semi-molten deep Earth could similarly move the thin, brittle rock of the crust.

Next, Griggs studied measurements of hundreds of earthquakes gathered by earthquake experts Beno Gutenberg and Charles Richter. The earthquakes were caused when two gigantic slabs of rock suddenly slipped past each other, with the break often starting many miles beneath the surface. Griggs seized on this work to suggest that perhaps the earthquakes were caused by movements of the rock down at the boundary between the crust and the *mantle* where these deep convection currents hit the crust.

However, despite these findings by a few bold, unconventional scientists, the overwhelming majority of geologists still dismissed the idea of galloping continents. Instead, they clung ferociously to the increasingly complicated efforts to explain the design of the Earth in terms of the shriveled apple. But World War II and the imminent discovery of the Mid-Atlantic Ridge soon changed everything. In 1939, Adolf Hitler unleashed his blitzkrieg assault with tanks and bombers, quickly conquering much of Europe. Soon a stunned Great Britain found itself facing the might of the Nazis across the narrow English Channel. Suddenly, Britain's survival depended on a lifeline of supplies from the United States. When Hitler realized he could not defeat the British fleet and invade England, he unleashed his U-boats to sink the ships on which Britain now depended.

The U.S. and British navies quickly realized that the war depended on finding and sinking the German submarines. As a result, the U.S. Navy poured money into anything that might help them navigate the deep ocean and locate lurking submarines. The flood of research money supported a handful of scientific labs that specialized in the study of the ocean, including Woods Hole Oceanographic Institution in Massachusetts, Scripps

WHERE GIANT SQUID LURK

The strange, ancient varieties of squid remain among the most fearsome and mysterious predators in the great dark expanse between the ocean's sunlit surface and the inky depths at the top of the Mid-Atlantic Ridge. The giant squid, a 50-foot- (15-m-) long creature that has inspired tales of sea monsters, has rarely been glimpsed, save as tattered remains washed ashore or pulled from the depths in fishing nets.

Closely related to shellfish and octopi, squid come in 181 species ranging in size from one inch (2.5) to 50 feet (15 m). They boast 10 arms, including two, long, sucker-tipped tentacles for snagging their prey. They jet about by squirting water out of their bodies and can change their color with a thought. They can outsprint the fastest fish for a short distance and also migrate thousands of miles.

They have enormous eyes and a gigantic brain, relative to their size. The squid has the largest nerve fibers on the planet, which makes possible a hair-trigger nervous system. The massive giant squid has never been captured alive, but a team of Japanese scientists recently used a robot camera to take the first picture of one, using a piece of bait dangling on a long line. The squid got tangled in the line and ripped off one of its enormous tentacles in its struggle to break free.

The giant squid probably fears nothing but the sperm whale, another miracle of nature. Biologists still do not know exactly how the sperm whale manages its hour-long dives to depths of a mile or more. Sperm whales can generate a sonic blast that perhaps stuns their prey. Biologists have found the six-inch- (15-cm-) long jaw-like beaks of giant squid in the stomachs of sperm whales killed by whalers, along with the scars of the squids' suckers on their skin.

Institution of Oceanography in California, and Lamont Geological Observatory at Columbia University. These three research laboratories would ultimately lead one of the history's scientific revolutions.

For instance, the navy immediately began probing the ocean with *sonar* sound waves, hoping to get an echo off a hidden submarine. But the captains soon discovered that the sonar did not work very well in the afternoon. They figured that the sonar operators were dozing after lunch or that mysterious swarms of underwater creatures were creating echo shadows. So the navy asked Maurice Ewing at Woods Hole to solve the mystery.

Ewing discovered that temperature changes below the surface affected the sound waves, creating a shadow in which a submarine could hide. So he devised a way to focus the sound waves to penetrate the shadow region. This had the added benefit of providing a much more accurate way to map the seafloor.

Next, scientists realized they could use these sound-generated maps of the seafloor to help thousands of ships navigate the Atlantic. Systematic mapping projects soon revealed mysterious, flat-topped mountains rising from the flat plain of the Atlantic. Scientists determined that these were former volcanic islands that had been drowned by the rising sea level, their tops flattened by wave action. The *seamounts*, or guyots, provided excellent navigational markers for ships crossing the ocean, but they also posed a new mystery for geologists. Many of these flat-topped mountains remained thousands of feet beneath the surface. Scientists did not understand how they got there and how they could have ever been close enough to the surface for waves to flatten their tops. Not even the most frigid of ice ages could have lowered the sea level so far.

The new sonar maps soon revealed a stunning view of the seafloor. The most striking feature was a 12,000-mile- (19,300-m-) long Mid-Atlantic Ridge—part of a chain of mountains that stretches for a total of 42,000 miles (67,600 m). The chain of underwater mountains is four times longer than the Andes, Rocky Mountains, and Himalayas combined.

The vast chain of mountains running nearly pole to pole down the middle of the Atlantic Ocean covers a mind-numbing one-quarter of the Earth's surface. Its bends and kinks mirror the outlines of both North and South America, Europe, and Africa. The steep, jagged zone of mountains rearing upward from the seafloor averages about 500 miles (805 km) in width, with many peaks rising 20,000 feet (6,100 m) from the seafloor. For most of its length, a narrow, mile-deep (1.6 km) *rift valley* snakes along the crest, deeper and wider than the Grand Canyon. Moreover, thousands of massive east-west canyons and ridges continually crack and offset the north-south Mid-Atlantic Ridge.

> ### PUERTO RICO TRENCH: DEEPEST ATLANTIC OCEAN TRENCH
>
> The warm Caribbean Sea is an arm of the Atlantic Ocean and contains the Atlantic's deepest place. The six-mile- (9.7-km-) deep trench is similar to the arch of islands that includes Antigua, the Virgin Islands, Puerto Rico, the Dominican Republic, and Haiti. Here a relatively narrow rift in the crust plunges to 28,232 feet (8,605 m) beneath the ocean's surface, nearly a mile shallower than the Mariana Trench in the Pacific Ocean, but still an awesome descent into darkness.
>
> The Puerto Rico Trench forms the boundary between the Atlantic Ocean and the rich Caribbean Sea. This small, warm sea between Cuba and Central and South America harbors a tropical profusion of sea life, a welter of islands, and the world's second-longest barrier reef.
>
> The geology of the Caribbean is even more complex. Here the North American, South American, and African *crustal plates* have collided, creating the trench, the chain of islands, four major ridges, three deep basins, and a network of deep *fissures* in the seafloor.

To add to the mystery, research ships had dragged iron dredges across the flanks of that stunning ridge. Those dredges brought to the surface raw chunks of young volcanic basalt. The emerging images of the largest mountain range on the planet would have amazed the world's geologists, who believed the seafloor was a flat, barren desert. However, almost all of the information came from navy ships and remained a wartime secret. These detailed maps of the seafloor gave American ships a huge advantage in navigation. They also gave Britain and the United States an edge in the ceaseless search for those lethal German submarines. However, even after the war ended, the military refused to let the scientists release the detailed maps.

MID-ATLANTIC RIDGE MAPPED

Scientists Bruce Heezen and Marie Tharp resolved to find a creative way to get around the security restrictions. They converted the detailed numbers into an exquisite map of the seafloor. They used color-coded shadings to exaggerate the vertical relief but left out the specific depths, which remained classified. The stunning image, released in the 1950s, transformed geologists' view of the seafloor. The image shows the massive Mid-Atlantic Ridge, which breaks the surface at volcanic Iceland. Throughout its twisting path, the Mid-Atlantic Ridge mirrors the outlines of both the Americas and the Old World, filling much of the space between the continents.

Even more dramatically, the Mid-Atlantic Ridge connects to a planetary system of similar ridges running down the middle of each of the world's major ocean basins. Moreover, the network of ridges is echoed on the opposite side of the ocean by a network of canyons or trenches, some

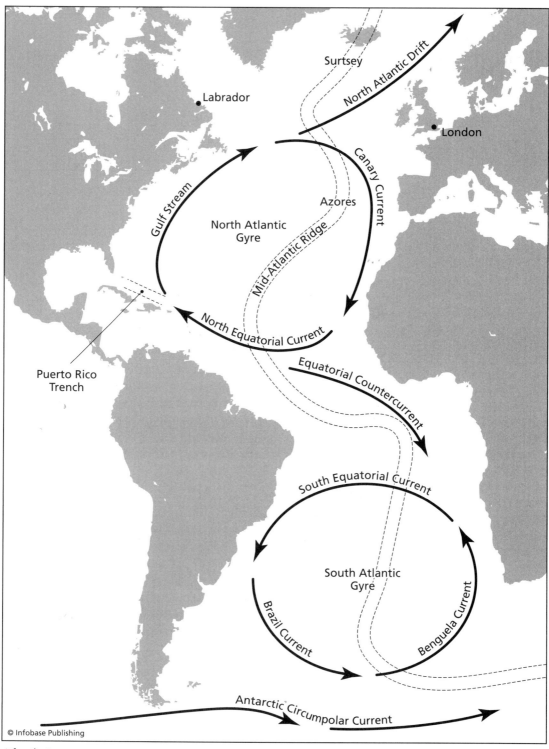

Atlantic Ocean

deep enough to swallow up Mount Everest so thoroughly that its peak would still be more than a mile beneath the surface of the ocean.

Clearly, the scientists mapping the ocean bottom had discovered a new and unexpected world. Geologists would have to abandon the notion of a flat, boring seafloor and explain a world every bit as dynamic, surprising, and active as the continents.

Everything had changed, but most geologists still did not understand the implications. Wegener remained an oddball with a strange theory. But not for much longer.

2

San Juan de Fuca Ridge
Pacific Ocean

The next act in the deep-sea revolution that would transform geology depended on lurking Nazi submarines, desperate young graduate students, a brilliant woman who could not get into graduate school, a splintered continent, and a missing chunk of the Earth. Geophysicist Tanya Atwater revolutionized the understanding of the Earth's structure when she unraveled a deep-mud mystery surrounding a nearly buried, underwater mountain chain off the western coast of North America, later known as the San Juan de Fuca Ridge. But first the lurking Nazi submarines and some deadly floating mines.

After Adolf Hitler crushed the French and conquered Europe, the British prepared for a long, terrible siege. Their survival now depended on supplies brought by cargo ships from the United States and its own scattered empire. Hitler did not have the navy he needed to strangle the British, but he did have both submarines and mines. He unleashed his U-boats, which quickly sank so many ships the British ran short of food.

Both American and British leaders realized that Hitler would win the war unless scientists figured out how to find submarines and protect ships from mines. So the Americans and the British both poured money into studying magnetism. They hoped that scientists could build very sensitive detectors that could sense the *magnetic field* created by a metal submarine hidden beneath the surface. The navy also wanted to find ways to prevent mines triggered by magnets from going off whenever they came near the metal hull of a ship. They knew that undersea mountains and masses of magnetic rock create tiny changes in the magnetic field of the Earth—little bumps of magnetic force that when graphed stand out like a sock under a bedspread. If they could measure and map those magnetic bumps, they could help ship captains leading convoys and destroyers figure out where submarines were beneath the featureless ocean surface.

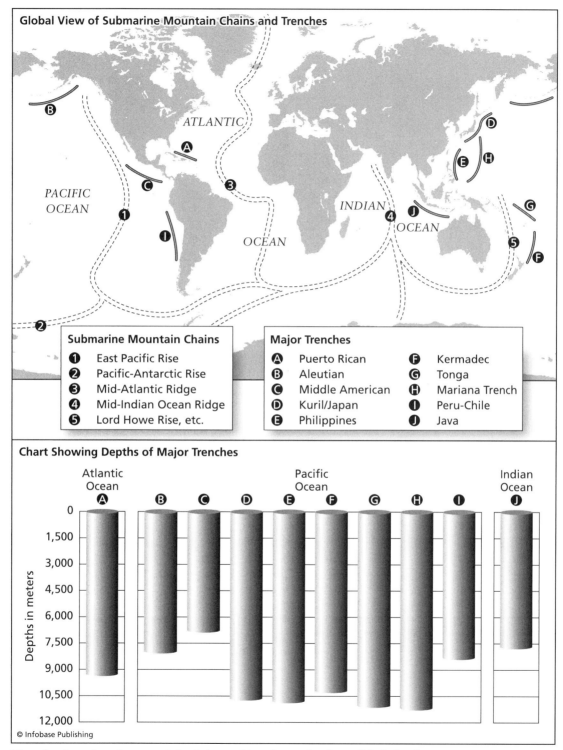

Map of major submarine mountain chains and trenches and a chart of their depths

So in order to win the war, the government invested heavily in solving one of the great mysteries of the Earth, the workings of the planet's magnetic field that makes compass needles point north; guides geese, sea turtles, and whales; and causes particles expelled from the Sun to spiral downward at the poles to create the ethereal "northern and southern lights."

Suddenly, scientists who had been scrounging for money for years found themselves drafted and put to work trying to defeat mines and submarines. The U.S. Navy quickly hired Teddy Bullard to figure out ways to locate and disarm mines. Bullard was one of the world's greatest experts on the Earth's baffling *magnetic field*. Bullard was also one of the scientists who helped prove that when *lava* or *magma* cools and hardens, tiny magnetic particles in the rock line up with the Earth's magnetic field and point north. That finding would later revolutionize our view of the Earth.

The scientists mobilized by the navy soon designed sensitive devices called *magnetometers* that could be towed behind ships to measure magnetic fields. The navy hoped these devices could detect submarines. But they also could measure the magnetic properties of the rocks in the ocean floor. They quickly discovered all kinds of strange magnetic variations in the rocks on the ocean bottom. Sometimes, they would find an undersea volcanic mountain with a strong magnetic field because it had lots of metallic rocks such as iron. Other times, they would find patterns that proved lava had spilled out to cover areas on the seafloor. As the lava cooled, the magnetic elements in the rock lined up with the Earth's magnetic field. These magnetic variations on the seafloor could be used to create detailed maps, so that the captain of a submerged submarine or a ship could figure out his ship's position based on the magnetic patterns in the seafloor. This prompted the navy to send ships throughout the oceans to measure the magnetic properties of the seafloor and so create a new kind of map.

PROBING THE MAGNETIC FIELD

After the war, some top scientists like Bullard deliberately moved away from military research, shaken by the role of scientists in developing the nuclear bomb. Bullard focused on the Earth's magnetic field and how it affected rocks. For instance, Bullard argued that the Earth had a *core* of molten iron and that vast currents in that core produced the planet's magnetic field. Another leading theory suggested the Earth's spin produced the magnetic field. To prove his theory, Bullard descended thousands of feet into coalmines to measure the Earth's magnetic field. If the spin of the Earth created the magnetic field, then it should get a little weaker far beneath the surface. But if the magnetic field came from currents in a molten core, it should not change. Sure enough, Bullard could find no difference in the magnetic field in the deepest mines, suggesting vast currents churned down in the Earth's molten core.

Now scientists all over the world began measuring the magnetic orientation of magma that had hardened into rock. They quickly came up with bewildering measurements. Strangely enough, these little, natural compasses in volcanic rocks pointed every which way, depending on their age and what continent they wound up on. Sometimes rocks side by side pointed different directions, as though the North Pole had suddenly become the South Pole. Moreover, on some continents, the magnetic elements in the rocks that were, say, 50 million years old, pointed north. But magnetic particles in rocks of the same age on another continent would point east. This did not make sense.

Scientists set to work on the mystery. As always, such questions drive scientists and scientific progress. First, some scientists proposed a strange idea. Maybe every so often the Earth's magnetic poles flip or reverse so that north becomes south. That would explain why magnetic particles in rocks of different ages pointed in different directions: It all depended on which way the poles pointed when those rocks cooled. In the 1920s, Japanese geophysicist Motonari Matuyama studied lava rocks in Japan and found evidence that 10,000 years ago the poles flipped. But the growing hostility between Japan and the West in those days before World War II prompted most scientists in Europe and the United States to largely ignore Matuyama's work.

In the 1950s, another scientist, Jan Hospers, made the same discovery by studying volcanic rocks in the strange island of Iceland, which fumes, sputters, and rattles with constant volcanic activity. That island rises implausibly from the sea right at the northern end of the great chain of mountains known as the Mid-Atlantic Ridge, which *sonar* maps had revealed during World War II.

MYSTERIOUS MAGNETIC STRIPES

The fresh evidence that the Earth's poles seemed to sometimes flip inspired scientists to look at other volcanic rocks. Eventually, they found evidence that during the Earth's long history the magnetic poles flip-flop repeatedly, but at unpredictable intervals. Sometimes the poles stay the same for millions of years. Sometimes they shift after a few thousand years. Scientists cannot fully account for the flip-flops, but they have found no evidence that one orientation is more likely than another: North is as good as south. Geologists have created a magnetic calendar from the magnetic flip-flops and used it to date the rocks themselves, providing what amounts to a "tape recording" of the Earth's history based on the dynamics of the Earth's core. Intriguingly, some experts believe the Earth's magnetic poles are in the process of flipping right now, since the strength of the magnetic field has declined about 10 to 15 percent since they started making detailed measurements.

Meanwhile, the U.S. Navy was giving money to scientists to conduct basic research, which tries to answer basic questions even if those answers do not lead directly to useful inventions or products. Two laboratories led the world in ocean research, Scripps on the West Coast and Lamont on the East Coast. Every time researchers from Lamont went to sea on a research project, they towed a magnetometer behind them to map the seafloor. Then they filed away the information for someone else to worry about. Scientists from Scripps often did the same thing.

One Scripps research ship chugging along above the San Juan de Fuca Ridge off the northwest coast of the United States came back with very strange readings. The Scripps researchers measured a bizarre, magnetic zebra-stripe pattern in the magnetized rocks under the thick layers of mud dumped on the seafloor by the Columbia River. The scientists who made the measurement published the magnetic map in a scientific journal, a peer-reviewed magazine that announces basic research findings after other scientists have continued their calculations. But no one knew what to make of the zebra stripes, and so no one paid much attention for the next four years.

Meanwhile, other scientists puzzled over how to use these magnetic measurements on the seafloor to understand the many strange observations that suggested the continents of the Earth somehow change positions. Perhaps matching up the magnetic orientation of rocks on the continents and comparing them to the record of pole flipping gleaned from the *magnetic stripes* on the seafloor could be used to prove that the continents do move, just as the ridiculed Wegener had suggested decades ago.

Several scientists, including Canadian geophysicist Lawrence Morley and Cambridge geophysicists Frederick Vine and Drummond Matthews, hit upon a big idea at the same time. They stared and stared at the new maps showing these gigantic mountain ranges running up the middle of nearly every ocean basin and wondered the following: Suppose those ridges marked great cracks in the Earth. Suppose that liquid rock constantly pushed up into those cracks, shouldering aside the relatively thin *crust*. Then suppose the continents are really made up of lighter rock, floating like icebergs in the dense rock of the ocean crust. That would solve many mysteries, including the matched fossils, rock types, and magnetic stripes. In that case, argued some scientists, the continents are not drifting through the crust, but the crust itself is splitting open, forced apart by magma from deep in the Earth. In that case, the surface of the Earth would be shaped not by the drifting of the continents but by the spreading of the seafloor. But how could scientists prove that?

EARTH'S POLES FLIPPING?

The Earth's poles may be ready to flip-flop. That fascinates scientists, but could spell disaster for geese, sea turtles, and skin cancer sufferers.

Validating *plate tectonics* depended heavily on the discovery that the Earth's magnetic poles sometimes reverse themselves so that compass needles point south instead of north. The orientation of the north magnetic pole is recorded in the alignment of magnetic particles in cooling magma. Therefore, magnetic stripes on both sides of undersea ridges proved that the crust is continuously manufactured at the ridges then pushed outward on a geological conveyor belt.

But geologists still do not know exactly why the poles flip. It has something to do with the *convection currents* flowing in the Earth's molten core, just as an electric dynamo can generate a magnetic field. For instance, just such an electric dynamo charges your car battery as you drive.

Now it looks like scientists will get to study one of these mysterious, almost random pole flips up close. New measurements suggest that the Earth's magnetic field has weakened by 10 or 15 percent in the past 150 years. No one is sure why that is happening. Most scientists think it has something to do with currents in the molten core of the planet, circulating in great *convection cells* like the water in a pan on a hot stove. The European Space Agency recently launched three new satellites called the Swarm to measure the apparently deteriorating magnetic field with new precision. Although the deterioration of the magnetic field has accelerated in recent decades, scientists think that it could take 2,000 years for north to become south.

The poles last flipped 780,000 years ago. The poles flip on average once every 500,000 years, but the intervals vary widely. For instance, when the dinosaurs were on Earth, the poles remained stable for 35 million years.

A pole flip could have dramatic and unpredictable consequences. For instance, most migrating birds can apparently sense the Earth's magnetic field, thanks to magnetic particles called magnetite that float in their brains. So can loggerhead turtles, which routinely make 8,000-mile (12,870-km) migrations across the Atlantic. So do salmon, whales, homing pigeons, honeybees, frogs, Zambian mole rats, and a host of other species. No one knows how a magnetic field reversal will affect such creatures.

Even more disquieting, some studies have found massive die-offs of microscopic creatures at the base of the ocean's food chain during periods when the magnetic pole flips. In addition, the Earth's magnetic field also shields the surface from electrically charged cosmic rays released from the Sun. The spiral of cosmic rays down into the magnetic field at the poles causes the astonishing northern lights. If the magnetic field becomes weak or erratic as it flips, cosmic rays could bombard the Earth's surface. That could cause a big increase in skin cancer and other problems.

PAPER MODELS AND CRACKS IN THE EARTH

At this critical moment, another brilliant scientist showed up, Canadian geologist J. Tuzo Wilson, a scientific showman with a genius for seeing patterns in the confusion of statistics and measurements. He started off studying islands in the Pacific, such as the chain of islands that constitutes Hawaii. He noticed something odd. The islands got older the farther they were from the East Pacific Rise, which runs down the eastern side of the Pacific Ocean. The *Alvin* (the first deep-sea submersible capable of carrying passengers) can be seen exploring the bottom of the Pacific Ocean in the color insert on page C-1 (top). Wilson suggested that the East Pacific Rise was another great crack in the Earth where irresistible currents in the molten core and the semi-molten *mantle* hit the brittle rock of the thin crust. That was not a dramatically new idea, but what he came up with next provided a huge step forward. He visualized precisely how these deep currents could boil up against the crust of the ocean and pile strain on a crack in the surface of the Earth beneath the East Pacific Rise. Wilson realized that the pressure would not create a single, 40,000-mile- (64,370-km-) long crack. Instead, that long crack would break up into hundreds of shorter sections, each section fractured by smaller, offsetting fault lines. So while the crust would pull apart along the crack itself, along those offsetting faults the two pieces of the Earth would slip past each other. That is exactly what scientists studying earthquakes had always thought. They had discovered many of these giant cracks in the Earth. But none of those cracks pulled apart as the people pushing for *seafloor spreading* insisted. Instead, great chunks of the Earth constantly slipped past one another.

Next, Wilson connected his theory to the *San Andreas Fault*, the most famous earthquake fault in the world. This massive, deadly crack in the Earth runs from the narrow Gulf of California, all the way up the coast of California, and then plunges into the ocean. As it turns out, it lines up quite nicely with the Juan de Fuca Ridge, hidden in the ocean off the coast. The two sides of the San Andreas Fault slip past each other: They do not pull apart like you would expect if they were part of one of the cracks in the Earth that spur the supposed seafloor spreading. But Wilson suggested that perhaps the San Andreas Fault is one of those offsetting *"transform" faults*. To the south, the spreading of the fault had opened up the Gulf of California. To the north, the spreading center of the San Juan de Fuca Ridge lay mostly buried under mud dumped on the ocean bottom by the gigantic Columbia River. So Wilson figured out how the various faults would slip and slide and spread and came up with some predictions that earthquake experts could actually test by measuring movements deep in the Earth that caused earthquakes.

THE SAN ANDREAS FAULT

The San Andreas Fault remains one of the best-studied, most baffling, most dangerous earthquake faults in the world. It starts near the head of the Gulf of California, runs nearly the length of California, then plunges into the ocean in the Pacific Northwest at Cape Mendocino. Coastal California on the west side of the fault is moving past the rest of California at about the speed a person's fingernails grow. But movement along the fault comes in great, destructive lurches instead of at a steady rate. Along most sections of the fault, the friction between the rocks on both sides of the fault holds them in place for centuries at a

(continues)

The San Andreas (transform) Fault

(continued)

time as the strain gradually builds. When the strain finally overcomes the resistance of the rocks, these sections suffer terrible earthquakes. On the other hand, some sections of the fault move constantly, generating small earthquakes every few years.

One of those sections of the fault that moves often and therefore produces frequent, small earthquakes lies near the California town of Parkfield. Small earthquakes take place once every 20 years or so in this section, but despite years of intensive study, scientists still cannot predict when the next quake will come even in this well-behaved stretch of the fault.

So they also remain helpless to predict the devastating quakes the fault produces once every century or so. The San Andreas has produced three massive earthquakes in relatively recent times, including the following: In 1857, a 8.0 quake broke loose a 220-mile- (354-km-) long section of the fault in central California, killing two people; in 1906, a 7.8 quake ruptured a 270-mile (435-km) section in northern California, devastating San Francisco and killing an estimated 3,000 people; in 1989, a 7.1 quake occurred along a 25-mile (40-km) section of the fault near Santa Cruz, California, killing 63 people.

Wilson came up with a theory that explained how the Earth could spread apart along mid-ocean ridges but still generate lots of the so-called *strike-slip faults* that earthquake experts had already documented all over the world. Wilson even developed ingenious paper models that showed how a spread-apart fault could create all these strike-slip faults. Wilson was a big hit at scientific conferences, especially when he pulled out his little, folded paper models to illustrate the complex mathematics of his talk.

Geophysicist Tanya Atwater finally pulled it all together. At one conference, she was fascinated by Wilson's lecture on *transform faults* like the San Andreas Fault, complete with paper models. At another conference, she listened to a talk about the strange, zebra-stripe pattern of magnetic stripes on the bottom of the ocean along the San Juan de Fuca Ridge, running like a northern extension of the San Andreas Fault. Those maps were based on measurements taken routinely then filed away and nearly forgotten. Now with the growing excitement about the new theory of *seafloor spreading* and Wilson's efforts to understand transform faults, some scientists were taking a second look at these strange, sea-bottom magnetic stripes.

Atwater remembers her excitement when she first saw the map of the magnetic stripes along the San Juan de Fuca Ridge. "It was like a bolt of lightning had struck me. My hair stood up on end. My sisters still remember how crazy I was at dinner that night. I was crazy-excited: This was the big picture key I had been dreaming of." She immediately changed her life plan and rushed to apply to study at the Scripps Institute of Oceanography, one of the leading centers for the study of the sea.

Scripps scientists had been routinely making measurements of the seafloor for decades, so they had a massive treasure trove of measurements that would finally prove or disprove this updated version of Wegner's long-ridiculed theory.

UNRAVELING THE MYSTERY

Atwater was soon consumed with the effort to explain the complicated forces pulling the Earth apart along these still mysterious ridges and trenches. She remembers standing in front of a wall-size map of the Earth in one lab on which someone had plotted the *epicenters* of thousands of earthquakes. The pinpoints of the epicenters formed a distinct pattern, clustered close to the surface along the mid-ocean ridges and miles beneath the surface in the undersea trenches.

She went to work trying to understand one of the most baffling and well-measured pieces of seafloor in the world, the area along the San Juan

---

A WOMAN JOINS THE BOYS' CLUB OF SCIENCE

The crucial task of validating the theory of plate tectonics fell finally to a brilliant young woman who almost could not study geology because graduate programs at the best universities even in the early 1960s would not accept women. Tanya Atwater wanted to be an artist until 1957, when the Soviet Union launched *Sputnik,* the first spacecraft to circle the Earth.

So she decided to become a scientist. But when she graduated from high school and applied to the internationally renowned California Institute of Technology, administrators said they would not admit her because she would no doubt just get married, quit science, and waste her expensive schooling. Harvard agreed and suggested she apply to Radcliffe, a woman's school. However, Radcliffe rejected her application because she did not take Greek or Latin in high school. Fortunately, her mom was a botanist and her dad an engineer, and their encouragement and example helped Atwater ignore those discouraging rejections.

Instead, she applied to the Massachusetts Institute of Technology, one of the few top science schools that would accept women in the late 1950s. She soon found herself drawn to geology by her love of the outdoors. She then attended graduate school at the University of California at Berkeley during the turbulent 1960s.

Even after she got her degree, she faced resistance in a male-dominated field. For instance, her research required weeks and even months at sea. But many sailors and oceangoing scientists thought a woman on a ship at sea brought bad luck. Later she discovered that her attempt to get on board a ship for a research trip often triggered bitter debates among the men. And when she did tour top labs with her colleagues, other scientists often ignored her or assumed she was someone's girlfriend.

The slight only strengthened her resolve to earn her place in science by working so hard that she eventually gained a much deeper understanding of how the Earth works than any of those guys who wondered whether they ought to let her on board the boat.

de Fuca Ridge. Here the San Andreas Fault plunges into the sea and a long ridge lies buried under mud. Along both sides of that buried ridge magnetic zebra stripes record the flipping of the Earth's poles. In addition, a network of earthquake faults connects to both the San Andreas Fault and that mostly buried ridge.

The ideal workings of science are revealed in Atwater's brave effort to combine all of these confusing clues into a single, elegant description that would prove or disprove a jumble of contending theories. Her struggle shows how science is supposed to work. She started out trying to understand theories developed by other scientists. Then she gathered up measurements taken in the real world. Finally, she worked herself into exhaustion trying to match the theories with the measurements. It proved revolutionary, a perfect illustration of the excitement and satisfaction that motivates the best scientists.

The Good Friday earthquake of 1963 in Alaska along deep fissures at the boundary between two crustal plates caused massive damage. *(National Oceanic and Atmospheric Administration)*

She combined all of these theories and measurements and in cooperation with several other scientists developed a brilliant description of the seafloor in the Pacific Northeast. That description proved crucial in converting most geologists to the theory of plate tectonics. This revolutionary theory explained drifting continents, the jigsaw puzzle continents, the matching fossils and rocks, the magnetic stripes, the pattern of earthquakes, and many other previously baffling observations.

The San Andreas Fault represents one of Wilson's enigmatic transform faults offsetting the spreading center crack in the Earth's crust that separates the Pacific and the North American *crustal plates*. The land on one side of the fault has moved north by some 300 miles (483 m) relative to the land on the other side.

Before the San Andreas Fault formed, the Pacific plate and the North American plate had simply smashed into one another. The resulting earthquakes and deep-down heating created the massive, granite bubble of the Sierra Nevada. But the bumping and grinding of the edges of those two vast plates shifted some 75 million years ago, creating the offsetting *San Andreas Fault*. That offset edge dividing two plates has since generated some of the planet's biggest earthquakes.

That collision of plates got even more complicated under the seafloor around the San Juan de Fuca Ridge. Atwater and her colleagues reconstructed the magnetic stripes and other measurements to conclude that three different plates collided here, forming a complex "triple junction" between the giant Pacific and North American plates, with the small,

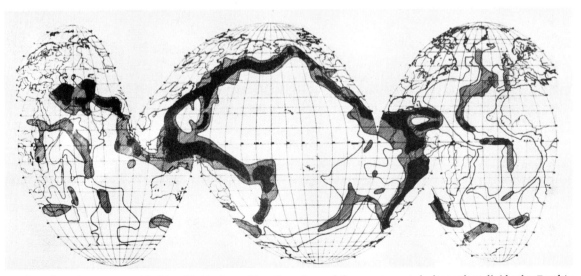

Earthquake epicenters worldwide effectively outline the edges of the great crustal plates that divide the Earth's surface. *(National Oceanic and Atmospheric Administration)*

MOUNT ST. HELENS

The same bumping and grinding of crustal plates that created the nearly buried San Juan de Fuca Ridge also generated one of the best-documented and dramatic volcanic explosions in modern history.

After months of fuming and fussing, Mount St. Helens in the Cascade Range of the Pacific Northwest erupted on May 18, 1980, blasting surrounding mountainsides with ash, followed by devastating mudslides.

The magma feeding Mount St. Helens and the other volcanoes that built the Cascades came from an ancient, melted-down crustal plate that had been crunched and buried just offshore where the San Juan de Fuca Ridge marks a rare triple-boundary crunch of crustal plates.

Magma had filled the vast chamber feeding smoke and ash out of the top of the cone-shaped volcano when a 5.1 Richter scale earthquake caused the mountains on the north face to collapse. As a result, lava and ash blasted out of the side of the mountain, almost instantly devastating 230 square miles (596 km²) of forest. In addition, a mushroom-shaped column of ash rose thousands of feet, spreading ash for thousands of square miles. The explosion collapsed the top 1,200 feet (366 m) of the mountain, and the resulting landslides and outpouring ash and lava buried a 24-square-mile (62-km²) valley and killed 57 people.

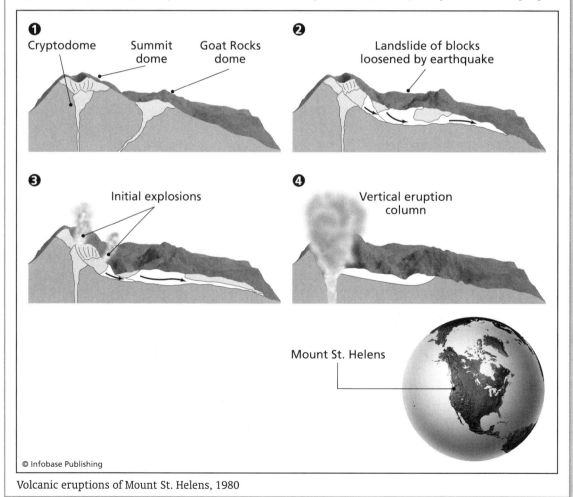

Volcanic eruptions of Mount St. Helens, 1980

fragmented Farallon plate caught in the crunch. They pulled together the work of other scientists and fit it into a larger pattern. They concluded that a once mighty crustal plate had been caught in this triple collision. Geologists still work to understand the complex results, both on the surface and miles beneath. Apparently, the Farallon and the small Kulu plate were spreading apart from one another when the much larger North American plate crashed into them. The larger plate with the light rock of North America embedded on top rode up and over the two smaller plates, like a truck smashing a sports car. These two buried plates were forced down toward the semi-molten mantle, and as the temperature and pressure built up, the rock was melted to form magma, which made its way upward to the surface to form a chain of arc *volcano*es that built the Cascades in the Pacific Northwest as well as the Aleutian Islands, off the coast of Alaska.

The 1970 paper by Atwater and her colleagues brought everything together. It explained the magnetic stripes, earthquakes, displaced fossils, disappearing plates, the Sierra Nevada, buried undersea mountains, fossils found in seafloor drilling cores, strike-slip faults, *spreading centers*, and subducted plates. It connected theories and observations by many other scientists. In the end, it explained seemingly disconnected data in a single, encompassing, logical, testable theory. And that demonstrates beautifully how science works.

PLATE TECTONICS TRIUMPHS

The masterful analysis of the San Juan de Fuca ridge system also all but settled the debate about plate tectonics. Scientists in the forefront of the new theory had accepted the basic ideas of plate tectonics for more than a decade. But most geologists worked on the continents where they could not easily see the effects of all this bashing, jostling, and burying of crustal plates. The rocks of the continents generally "float" on the denser rock of the plates. Without the discovery of mid-ocean ridges and trenches, geologists might not have figured out the Earth's surface is divided into these gigantic plates. They would have continued to refine complex, incorrect theories about how mountains formed, without seeing the underlying pattern.

However, geologists studying the seafloor had in just 30 years set loose a scientific avalanche of theories and evidence that buried the long-established theories first generated by the geologists laboriously studying the rocks of the continents.

Almost overnight, geologists abandoned all their old theories and embraced plate tectonics. Like any good scientific theory, the new way of connecting things solved many baffling mysteries. In the end, plate tectonics proved one of the most sudden, important, and dramatic scientific revolutions in human history—and all because of a German weatherman killing time in a library, a desperate effort to sink German submarines, the tedious collection of data by towing magnetometers, a woman who insisted on doing science, and the willingness of many scientists to ignore the conventional wisdom to take a chance on a new idea.

3

The Mariana Trench
Pacific Ocean

Crammed into a three-piece steel sphere with five-inch- (13-cm-) thick walls, Jacques Piccard and Lieutenant Don Walsh plunged into the world's deepest place at an inexorable three feet (.9 m) per second. Already deeper than any human beings had ever gone, they could feel the penetrating chill of the 32,500 feet (9,900 m) of ink-black waters straining to crush the ungainly craft in which they hoped to make history. Their

Trieste in drydock

tiny steel capsule, designed to resist the crushing pressure of a column of water seven miles high, hung suspended beneath a great, 50-foot- (15-m) long steel vessel. It was 1960, and the revolutionary *submersible Trieste*, as shown in the color insert on page C-1 (bottom), was descending toward the world's deepest place, an undersea counterpoint to Mount Everest called the Mariana Trench. The mile-wide pit of *Challenger Deep* lay in the bottom of the Mariana Trench near the Pacific island of Guam. If Piccard and Walsh could descend to the mysterious, 35,700-foot- (10,880-m-) deep bottom of Challenger Deep, they would achieve one of the great milestones of human exploration.

Challenger Deep is a pit in the bottom of an intermittent gouge of miles-deep canyons and trenches that run along the eastern edge of the Pacific. The jagged line of trenches could easily swallow the Grand Canyon. While the average depth of the Pacific is about 12,000 feet (4,000 m), the trenches descend to more than 35,000 feet (10,670 m) below sea level. To the east of these trenches rise chains of volcanic islands, including New Guinea, the Philippines, the Mariana Islands, Japan, and the Aleutian Islands. The system of trenches runs for some 30,000 miles (48,280 km). A mind-numbing 70 percent of the Earth's earthquakes are concentrated in that zone.

Back in 1960 when Piccard and Walsh made their historic dive, no one fully understood what had created the trenches, island arches paralleling the trenches, or nearby earthquakes. The sea bottom remained a mysterious place, unvisited and only dimly imagined. But roughly an hour into their descent, Piccard's and Walsh's imaginations got a nasty jolt.

Suddenly, they heard a terrifying explosion somewhere outside the steel sphere that protected them from the crushing nine tons (8 metric tons) per square inch of pressure exerted by the dark, frigid water overhead. They wondered then what had collapsed above them in the steel-hulled blimp loaded with some 28,000 gallons (106,000 L) of gasoline. Lighter than seawater, the gasoline's natural buoyancy would carry them back to the surface after they dropped the metal weights now bearing them down to the unknown bottom. They knew that the only way they could get out of their chamber when they returned to the surface was to climb through a long escape hatch running through the gasoline-filled hull. Had the tube collapsed? Had the hull ruptured?

Piccard and Walsh stared at one another, wide-eyed and silent as they waited for something else to happen—a lurch, a groan, a collapse, a plunge to death. But nothing happened. Instead, the *Trieste* continued to fall silently through the blackness that gathered with crushing force. They stared at the depth gauge, the needle dropping without hesitation or pause. Soon, it read 36,000 feet (10,970 m) below sea level. Impossible, they thought. That would put them 200 feet (61 m) deeper than

> **PACIFIC OCEAN: VITAL STATISTICS**
>
> - could swallow up an area equal to all the continents or two Atlantics
> - has a surface area of 60 million square miles (155 million km^2)
> - averages 14,000 feet (4,270 m) in depth
> - has a series of 35,000-feet- (10,670-m-) deep trenches
> - the Mariana Trench runs between the Pacific plate and the Philippine plate

the lowest point in the ocean, as previously measured by ships on the surface. Surface ships had carefully measured the depth of this deepest of places by setting off some 800 depth charges and listening to the *sonar* echo of those sound waves bouncing off the bottom. But the *Trieste* continued its seemingly impossible descent. Soon, the depth gauge suggested the damaged craft had dropped past 37,000 feet (11,280 m), which was below the seafloor as measured from the surface.

HISTORY OF A DREAM

As he descended into history and peril, Jacques Piccard must have wondered at his fate in his brave attempt to fulfill his father's dreams, a seven-mile (11-km) descent that marked an epic of human exploration.

The dream had started with his father, Auguste Piccard, a daring scientific adventurer who first made his mark using a balloon to rise higher into the atmosphere than anyone had ever gone.

The old technology for exploring the deep oceans had reached its limits in 1949 when Otis Barton reached a depth of 4,500 feet (1,370 m), about half the average depth of the ocean. But the weight of the cables attached to the steel sphere limited the whole endeavor. Besides, if the long uncoiling cables snapped or tangled, the crew in the heavy steel sphere far below the surface would be doomed, with no way to return to the surface.

Auguste Piccard proposed a revolutionary new design, based on the model of a hot-air balloon. He would free the pressure-resistant sphere from the tangle of cables connected to the support ship. Instead, he would sling the heavy steel sphere underneath an undersea "balloon." The challenge would be to find a way to protect the crew from the crushing weight of the ocean. Even nuclear submarines cannot get to the bottom of the ocean, because the hull would have to be massively thick to protect all of the living space inside. So Piccard resolved to design a steel "gondola" with thick steel walls just barely big enough for two crewmembers. He then designed a much larger, fluid-filled balloon structure so the submersible could both sink and rise without the aid of

> ## AUGUSTE PICCARD (1884–1962)
>
> Born in Switzerland on January 28, 1884, Auguste Piccard at first seemed a conventional scientist. He earned a Ph.D. in physics from the Swiss Institute of Technology and turned his attention to the study of cosmic rays, elemental particles expelled by the Sun's furious nuclear fusion reaction. He worked for a time with Albert Einstein on cosmic rays, as Einstein pondered mysteries of light and gravity to derive the formulas that would revolutionize physics. Piccard focused on strange bursts of electricity measured in the upper atmosphere that physicists thought might be caused by cosmic rays from the Sun skittering into the Earth's *magnetic field*. But Earth's thick atmosphere absorbed almost all of the Sun's cosmic rays, which was good for Earth-bound sunbathers but very bad for cosmic ray–loving physicists.
>
> Piccard's audacious solution was to design a balloon that would carry him into the upper atmosphere where he could better study his beloved cosmic rays. So he and an assistant climbed into a small basket hung beneath a hydrogen-filled balloon in 1931, determined to go higher than any human being in history. On that first run, they almost succeeded too well. The balloon shot up to 50,000 feet (15,240 m) in 28 minutes, with Piccard and his assistant packed into a pressurized aluminum gondola beneath the balloon. But just as they reached the blue-black, upper 10 percent of the atmosphere, they sprang a leak in their oxygen supply. Piccard plugged the hole but then discovered that the rope connected to the gas-release value that would let them descend was tangled. They barely managed to untangle the rope, descending in the darkness to land precariously on top of a *glacier* in the Alps.
>
> Such a close call might have discouraged a sensible man, but Piccard tried again the following summer, this time along with his brother, an engineer. They rose to 55,800 feet (17,008 m), the highest ascent in history to that point, and returned safely to Earth.
>
> Soon Piccard was seized by the idea of turning his high-flying balloon inside out and creating an undersea balloon that would enable him to go deeper than any human being had ever gone. In fact, he had been fascinated by the mysteries of the deep ocean for most of his life, ever since reading about trawlers that brought to the surface in their nets bizarre, glow-in-the-dark fish dredged up from miles below the surface.
>
> Piccard died on March 24, 1962, in Lausanne, Switzerland.

cables attached to a surface ship. The larger steel hull would be filled with fluid instead of air, which means the pressure would be the same inside and out. That is why fish can swim about happily five or six miles (8 or 10 km) below the surface, because the pressure on the inside is the same as the pressure on the outside. These deep-sea fish do not have the air-filled swim bladders that most fish that live near the surface use to float in the water without effort.

Piccard's deep-sea dreams lurched toward reality in 1937 during an audience with Belgium's King Leopold III, who asked him about his research, expecting some colorful stories about *helium* balloons. Instead, Piccard mentioned his design for a submersible "balloon." The king was intrigued and promised his support. Galvanized, Piccard lined up additional support from the French.

Just when he was ready to start building, World War II broke out. The Nazis rolled into Belgium, King Leopold fled, and Piccard's plans were put on hold for the six-year horror of World War II. However, Piccard was a stubborn dreamer. As soon as the war ended, he resumed work and in 1948 completed his 22-foot- (7-m-) long bathyscaph, or "deep boat." The blimplike upper hull was filled with gasoline and attached to a steel sphere for the crew. He also attached weights to the *submersible*, which the crew could drop to return to the surface. To descend, they could flood empty tanks with water and then release *buoyant* gasoline.

Piccard's vision confronted numerous challenges. He had to cobble together funding for what seemed more adventure than science. His first unmanned test design was plagued by problems. Although the unmanned vessel reached a record depth of more than 4,500 feet (1,500 m), he had to jettison gasoline so that the support ship could lift the little vessel out of the heavy seas. Unfortunately, he then did not have enough gasoline to make a manned dive the next day, which had been the point of the expedition. The newspapers ridiculed his effort, ignoring the record depth achieved and focusing on the lack of a manned test. The negative reception helped dry up Piccard's funding.

However, the flawed test run did catch the attention of a young French navel officer named Jacques Cousteau, the charismatic, ambitious, and well-regarded coinventor of the scuba technology that allowed divers to breath from pressurized oxygen in tanks on their back instead of through hoses connected to a surface ship. Cousteau got the French government interested. However, the French wanted control of the project. The proud Piccard balked. But when his other funding ran out, the French simply bought out the Belgian government's investment in the project and took control of Piccard's sturdy, experimental craft.

THE PACIFIC BASIN

The Pacific Ocean is unique in that it features a vast, central depression dubbed the Central Pacific Trough, extending from the Aleutian Islands to the Antarctic, and from Japan east to North America. All the other *ocean basins* have spreading center ridges running down the middle, like the vast Mid-Atlantic Ridge, but the Pacific covers a gigantic crustal plate, bordered by ridges and trenches. The eastern half of the Pacific plate is being forced down under the North American plate. This causes the earthquakes of the *San Andreas Fault*, the rise of the Andes in Peru, and the Sierras in California. It also drives the *volcano*es that built the Cascades in the Pacific Northwest. The great majority of the world's volcanoes and earthquakes take place around the edge of this single, vast crustal plate.

Frustrated at losing control of his own invention, the determined Piccard sought financing for a new design. With money from Swiss and Italian investors, plus the backing of the Italian city of Trieste, in 1953 Piccard designed a new submersible. A German arms manufacturer found a way to manufacture a three-piece metal sphere with a diameter of about seven feet. The steel walls were four feet (1.2 m) thick, giving it a weight of 10 tons (9 metric tons). Perhaps the most striking innovation lay in the portholes. Previous submersibles used thick portholes of clear quartz. Unfortunately, the quartz windows could shatter if the enormous pressure at the ocean bottom came to bear on even a small scratch. So Piccard instead used six-inch- (15-cm-) thick, Plexiglas windows. The cone-shaped windows were 16 inches (41 cm) wide at the top and four inches (10 cm) wide at the bottom. As a result, the mounting pressure would squish the flexible Plexiglas, making it fit more tightly under the 18,000 pounds (8,165 kg) per square inch of pressure in the trench bottoms. He filled the 50-foot- (15-m-) long hull with gasoline, which would make the unwieldy craft float like a cork.

He also devised an ingenious solution to the problem of making the craft sink. On each side of the sphere that protected the crew, he attached chambers filled with metal pellets. Inside of that chamber he also attached enormous *electromagnets*. When he turned on the magnets, they would fuse the pellets into a single mass. But when he turned off the magnets, the pellets would separate and begin trickling out of the chambers, making the submersible lighter. This would give the craft's pilot precise control over the *Trieste*'s buoyancy.

In addition, Piccard rigged the two enormous batteries that drove the electrical equipment and the lights on the submersible to drop off the hull if the electricity ever went off. That would instantly lighten the *Trieste* by 2,500 pounds (1,134 kg), so that even in a dire emergency it would float to the surface rather than sink to the bottom.

However, Piccard soon found himself in a race with the French, who had modified and improved his original submersible. Both teams were racing to reach the deepest place in the ocean. They knew that the first team to the bottom of the world would reap the glory, and few would remember the names of whoever got there second. After all, most people recognize the names of Sir Edmund Hillary and Tenzing Norgay, who made it first to the top of Mount Everest, and Neil Armstrong, the first to set foot on the Moon. Few recall the second people to achieve these feats.

The competition heated up in a succession of dives. The French set the first record with a dive to 6,930 feet (2,112 m), a few hundred feet short of the bottom of the Mediterranean. After their depth finder failed, the French crew had stopped just short of the bottom for fear of crashing into the seafloor. Piccard topped them a few months later, descending to 10,390 feet (3,170 m) but nearly burying himself in the soft mud at the bottom of the Mediterranean. After that, he added an echo sounder to warn the pilot when the bottom was approaching.

Now the French group ventured out into the storm-tossed Atlantic Ocean, where the average depth far exceeded the 10,000-foot (3,048-m) bottom of the enclosed Mediterranean. Here they faced the vexing problem of towing the unwieldy submersible far out to sea to reach the deeper waters. Designed to move gracefully up and down miles beneath the surface, the blimp-like submersible at the end of a long towline endured relentless battering by the turbulent seas of the Atlantic. Nonetheless, the French set a new record in 1948. They descended gingerly in the bathyscaph. At a depth of 3,280 feet (1,000 m), an oil line in the capsule sprang a leak, spraying a fine mist of oil. Still, they continued to descend. After a nearly two-hour descent, they reached the bottom at 13,287 feet (4,050 m). They found a flat, sandy expanse, with ripples in the sand demonstrating the existence of powerful currents even three miles beneath the surface. They spotted beds of unearthly sea anemones that they described as "tulips of crystal." At one point, a nearly seven-foot- (2-m-) long shark swam lazily past the viewpoint.

But after only 40 minutes on the bottom, the lights suddenly went out and a shudder rumbled through the craft. A blown fuse had cut off the power, which had triggered the automatic release of the heavy batteries attached to the hull. So the craft rose unceremoniously to the surface, demonstrating the inherent safety of Piccard's bold design, even in the hands of his scientific rivals.

In the meantime, Piccard was stuck in the shallow Mediterranean by a lack of funding. He went to the United States, where he won the backing of Robert Dietz, a marine geologist who would later play a leading role in the development of the theory of *plate tectonics.* He also won over scientists at Scripps Institution of Oceanography in La Jolla, California, and Woods Hole Oceanographic Institution. Once the top U.S. oceanographers made their pleas for a craft that could reach the bottom of the deepest oceans, Piccard won backing from the U.S. government. In 1957, the Office of Naval Research agreed to fund a series of test dives.

Piccard had sought official backing for nearly 20 years, sustained largely by his almost obsessive determination. He had also enlisted his son in his long quest. So now Jacques Piccard took over much of the negotiations and piloting duties. Finally, the Piccards' timing was perfect. The cold war between the United States and the Soviet Union was heating up, and the two superpowers were engaged in an expensive scientific competition. In 1957, the Soviet Union launched its *Sputnik* satellite, stunning U.S. scientists and alarming defense officials. It was no coincidence that the United States in 1957 also funded 26 test dives by the *Trieste* to conduct scientific research in the depths of the Mediterranean.

The navy's Office of Naval Research offered to buy the *Trieste* and fund further dives, providing Piccard agreed to move his operations to San Diego. Piccard pondered the offer. He knew that he would never reach the true depths of the ocean without the backing of the United States, but he also knew that accepting its backing would cost him control of his precious submersible. In the end, he had little choice. He knew that the first human beings to go to the deepest place in the planet would earn fame and a spot in scientific history. If he did not accept American money, the French or even the Soviets would surely beat him to the bottom. So he agreed to surrender effective control of the *Trieste* to the Americans. However, he inserted into the contract a provision that gave him the right to pilot the craft on any dives that presented "special problems."

The scientists and engineers immediately set to work improving the *Trieste* for dives to below 35,000 feet (10,668 m). Soon the engineers finished the new, stronger crew chambers and expanded the "blimp" portion of the hull to hold an additional 6,000 gallons (22,700 L) of gasoline. To test the updated *Trieste*'s capabilities, they dove to 23,000 feet (7,010 m) in a deep spot of a seafloor trench near Guam. The dive in 1959 stole the depth record away from the French, who were still using their own updated version of Piccard's original design.

Finally, the federal funders authorized the dive into the Challenger Deep, the Holy Grail of deep-sea explorers. But to Piccard's dismay, they authorized only one dive. The Americans selected Lieutenant Don Walsh, the naval officer in charge of the submersible, and Andreas Rechnitzer, scientific head of the deep-sea program. For a moment, it seemed that despite his two decades of pioneering effort Piccard himself would be left behind while others realized the dream that had dominated the life of both him and his father.

Then Piccard remembered his contract. Surely, this dive presented "special problems." To the consternation of the Americans, Piccard invoked the contract and insisted on a seat on the historic, dangerous expedition.

INTO THE DEEP

Unfortunately, troubles started accumulating well before that frightening explosion as they neared the bottom of Challenger Deep. The enormous gasoline-filled blimp hull of the *Trieste* could not be hoisted out of the water and set on the deck of a larger ship. So the support ship had to tow the *Trieste* from Guam out to the Mariana Trench. With winter storms whipping up 25-foot (7.6-m) waves, the *Trieste* labored and waddled through the punishing swells, straining at the tow cable.

They reached the spot above Challenger Deep in heavy seas. Piccard and Walsh faced the daunting task of reaching the *Trieste* in the pounding waves and then getting onto the low-riding deck without being drowned or pounded to a pulp. Imagine trying to jump from a small boat onto the curved metal hull of a submarine as the swells lift the boat by 25 feet in a moment.

Somehow, they leaped safely onto the *Trieste* and made their way to a small "conning tower," built around the hatch and metal tube that led down through the gasoline-filled hull to the submersible slung beneath. The conning tower primarily served to keep the ocean from washing into the open hatch when the crew entered and left but also provided a place to attach various instruments.

Piccard and Walsh got their first bad news when they examined some of those instruments. They discovered the waves had ripped away the telephone that would allow them to talk to people on the surface after they were sealed up in the crew compartment beneath the surface. An instrument that measured the speed of their descent was also broken, which means they could easily crash into the bottom. Finally, an instrument for measuring vertical currents had been beaten to a pulp, which would make it harder to maneuver. Moreover, they would now have trouble measuring any possible currents on the bottom, one of their key scientific goals.

They had to make a fateful decision. They faced a 14-hour dive to make the 14-mile (23-km) round trip to the bottom and back. If they waited to repair the instruments, they might not return to the surface before dark, and the support crew on the surface might have trouble finding them in the dark. But Piccard would not be stopped now. He ordered the dive without waiting to repair the instruments.

Even after they started the descent, problems continued to mount. First, they hit the density barrier of the *thermocline*. Cold water is heavier and denser than warm water. Near the surface, wave action thoroughly mixes the water, resulting in uniform temperatures. But deep water is colder, denser, and less mixed. The thermocline is a temperature boundary where the water stops mixing and chills. Mild thermoclines can form in the bottom of a swimming pool and chill a swimmer diving to the bottom. Above Challenger Deep, the *Trieste* hit the first thermocline at 340 feet (104 m) down. The submersible all but bounced off the layer of cold, dense water.

Now the *Trieste* confronted the drawback of its own ingenious buoyancy system. The descent was controlled by the balance between the buoyancy of its 28,000 gallons (106,000 L) of gasoline and the pellet ballast gripped by the electromagnets. When the *Trieste* hit the cold, dense water at the thermocline, it made the warm gasoline in the hull comparatively more buoyant, abruptly halting the descent. So now Walsh had two choices: He could release some of the gasoline, so the *Trieste* would resume sinking, which would reduce his buoyancy safety margin when he wanted to return to the surface. Or, he could wait at the thermocline as the gasoline in the hull above cooled. As the gasoline cooled, it became less buoyant. At some point, the *Trieste* would start sinking again. However, that process also reduced the ship's buoyancy and used up precious time. If they returned to the surface after dark, they would have to float helplessly and wait for rescue crews, since *Trieste*'s tiny, battery-powered motor would be useless in fighting the battering swells at the surface.

Deciding that delay posed the greater risk, Piccard jettisoned some gasoline and the *Trieste* resumed its descent. It hit several weaker thermoclines during, the descent, each of which cost Piccard precious time and gasoline.

After breaking through the thermoclines, the long fall through the darkness went smoothly. They saw few signs of life beyond the Plexiglas

---

THE MYSTERY OF SEAMOUNTS

The oceans of the world hide one of the planet's most common landforms—isolated, dead, undersea volcanic hills called seamounts. Some of those dead undersea volcanoes have flat tops, thought to have been created by the pounding of waves. Geologists call these flat-topped seamounts guyots. They are especially common in the Pacific, where they rise from the dark ocean floor and provide a refuge for sea life. Moreover, many guyots are thousands of feet beneath the surface. How on Earth could waves shave off the top of an extinct volcano thousands of feet beneath the surface? Even when a severe *ice age* locks up huge quantities of water into the ice caps, sea level only changes by a few hundred feet. The embrace of the theory of plate tectonics provided the solution to the riddle of the guyots. Geologist Harry Hess, one of the key advocates of the theory of plate tectonics, suggested that these flat-topped undersea guyots were once volcanic islands lying close to one of the great spreading center cracks along one of the *crustal plates*. Near the swell in the *crust* along a mid-ocean ridge, these enigmatic islands broke the surface. But as they moved away from the spreading center on the conveyor belt of the plate, they stopped erupting and started sinking. Millions of years later, they were in the middle of the ocean far from the spreading center but still boasting a wave-sheared top. One of the highest mountains on Earth is a 26,650-foot- (8,123-m-) high seamount deep underwater on the edge of the seven-mile- (11.2-km-) deep Tonga-Kermadec Trench. It leans toward the trench and will eventually descend "down into the jaw-crusher," as Hess puts it.

viewpoints. Piccard fretted about whether now undetected vertical currents might be carrying them away from the mile-wide opening into the Challenger Deep as they fell through six miles (10 km) of water at a rate of three feet (1 m) per second. He also wondered whether they might drift into the steep, jagged walls of the trench itself, since they also could not accurately gauge the speed of their descent.

TERRIBLE EXPLOSION RESOUNDS

Piccard had just begun to feel optimistic when that terrible explosion resounded through the five-inch- (12.7-cm-) thick steel walls of the sphere that protected their frail bodies from a crushing pressure of some 16,000 pounds (7,260 kg) per square inch.

Walsh and Piccard tried not to let their minds sink into the grim possibilities as the *Trieste* continued to descend for an hour and a half after that still mysterious explosion. They could only shake their heads at the depth gauge that insisted they were already deeper than any of the surface measurements had indicated. Had they somehow fallen into a small, unsuspected hole in the bottom of Challenger Deep? Would they suddenly slam into hull-puncturing rocks in this dark tomb?

Nearly five hours after they had left the surface, the primitive sonar probing for the bottom cast back a black echo, warning that the bottom lay 252 feet (77 m) below. Piccard immediately adjusted the electromagnets to release a trickle of metal pellets, slowing the still not accurately measured descent rate.

Finally, the ooze of the bottom came into view in the dim spotlight through the thick viewports. A moment later, the *Trieste* touched gently down in the deepest place on Earth. The depth meter read an impossible 37,800 feet (11,520 m), some 2,000 feet (610 m) deeper than they had expected. Only later would they discover that the depth gauge had been calibrated in Switzerland in freshwater instead of salt water, which affected the measurements. Their true depth was later calculated at 35,800 feet (10,900 m).

Immediately, the spotlight fell upon a large, flat fish lying on the bottom. The sole-like fish gaped at them with two, huge round eyes. Humans and fish regarded one another with perhaps equal astonishment. Had this fish lived its whole life in these lightless, crushing depths? What must it have made of this apparition, gigantic and brilliantly lit? And why did a creature that lived seven miles from the memory of light even have eyes, gazing now at these alien intruders with lidless, implacable calm?

A moment later, the fish rose from the bottom and undulated nonchalantly off into the darkness. Piccard felt a great, unrestrained surge of joy, to have come to this place and found life and an escalation of mystery. He had achieved his great goal and would live to speak of it.

The practical, crisis-trained Walsh was not so sure. Craning his neck to look out his portal, he looked up toward the hull for some clue as to the source of that mysterious explosion. To his dismay, he could see a large crack in a plastic window in an antechamber that separated their pressurized sphere from the water-filled tube leading through the hull to the surface. They had to be able to seal that antechamber and pump out the water in order to open the hatch to their sphere and enter the escape chamber. But if the cracked window on the antechamber collapsed, they could not pump out the chamber and escape to the surface. If the surface crews could not figure out some way to release them, they might have to endure the pounding, dangerous tow all the way back to Guam still sealed in their steel coffin.

But Piccard shrugged and put that possibility out of his mind. Nothing he could do about it now. Instead, he stared rapturously out the window at the bland ooze that covered the very bottom of the world.

The seafloor was nearly featureless. The vanished fish now seemed more like some strange dream than a real creature. What had drawn that fish to such a desert, graced by barely a trace of the nutrients that had bloomed in the distant sunlight and sunk to this deepest place? In the Mediterranean, the seafloor was covered with mounds and burrows and tracks of living things. Here there seemed hardly a trace of life. They noted one parallel set of marks in the muck, perhaps the tracks of some unseen creature. They saw none of the ripples indicating a bottom current that they had seen even in other deep places. But they had only about 20 minutes to ponder the mysteries of Challenger Deep once they had finally arrived.

Unknown challenges lay between them and the surface, and time was slipping away. Piccard reluctantly began dropping ballast, and the *Trieste* rose again from the seafloor. Its speed rose soon to five feet (1.5 m) per second, a little faster than an elevator. However, after hours of uneventful ascent, the *Trieste* again hit the succession of thermoclines, this time from the underside. The eight hours in the near-freezing chill of the deep ocean had cooled the gasoline in the hull, so that it was now not much more buoyant than the warmer water near the surface. Once again, the *Trieste* struggled through the temperature barriers as its buoyant gasoline slowly warmed.

They broke finally into the brilliant sunlight at about 5 P.M., some nine hours after they had started. Breathlessly, they waited until the surface crew confirmed that the antechamber porthole had held.

They had gone deeper than any human being before or since and returned to tell the tale. No one has gone back to Challenger Deep since, nor systematically explored other trenches with similar depths. In 1997, a Japanese robot, Kaiko, briefly explored Challenger Deep, tethered by a cable to a ship on the surface.

Despite the triumph of Piccard's plunge, *Trieste* was already nearly obsolete by the time he returned to the surface. Unwieldy and expensive to operate, it also could barely move around once it reached the bottom. It had remained unchallenged in its ability to take human beings to the deepest places. But a new generation of much more flexible, compact submersibles would soon give scientists much greater flexibility and ease of access down to perhaps 20,000 feet (6,100 m), which would enable them to visit 98 percent of the ocean floor, including the much more active ridge systems.

Moreover, Piccard later realized that he and Walsh were lucky to have survived their daring adventure. Engineers calculated that the three-piece steel sphere on which Piccard's life had depended could easily have ruptured, filling with water and crushing the crew. The navy retired the *Trieste* in 1961 and never built a successor, so that no craft operating today can return to that depth. But Piccard had realized his great dream. The epic of daring had fired the public imagination, challenged a new generation of undersea researchers, and thoroughly perplexed one very odd fish. Now research would shift to the more accessible and dynamic undersea ridges to confirm the revolutionary theory of plate tectonics and reveal new mysteries.

4

The Galápagos Rift
Pacific Ocean

The discovery of the Mid-Atlantic Ridge and the analysis of the San Juan de Fuca Ridge system had largely vindicated advocates of *plate tectonics*, but they still faced one vexing mystery before the theory made sense. The theory suggested that, miles beneath the surface, molten rock was constantly squeezing upward and oozing into the ocean along giant cracks in the surface. The upwelling *magma* forced the *crustal plates* apart and created trenches and ridges. Now that scientists had mapped the seafloor, the theory made much more sense. But they still faced a problem. All that molten rock forcing its way toward the surface should have generated a lot of heat. But so far no one had found any trace of the *lava, volcano*es, and hot springs such titanic forces should produce.

The first deep-sea dive to visit a mid-ocean ridge took place in the early 1970s when a small fleet of tiny research submarines, or *submersibles*, dived to the Mid-Atlantic Ridge. They documented a bizarre volcanic landscape but found no trace of the missing heat. The results baffled the geochemists trying to explain the evolution of the ridges. The scientists believed that seawater filtering down through cracks in the rocks to those hidden magma chambers with molten rock heated to 2,552°F (1,400°C) should have produced some measurable increase in the temperature of the freezing, high-pressure water somewhere along the ridge.

The missing heat prompted some geologists and physicists at the time to question the whole theory of plate tectonics. Several theoretical models suggested that the vast, rising swell in the surface of the Earth on each side of the Mid-Atlantic Ridge was caused by the heat of the hidden magma miles beneath the surface, a great heat blister on the face of the Earth. But some of that hidden heat should be escaping into the ocean itself. The failure to find any hot springs or active lava flows prompted some scientists to challenge the newly triumphant theory of plate tectonics and suggest some other force had caused these baffling, awesome ridges of rock.

The mystery of the missing heat prompted another band of adventurous scientists to turn their attention to a completely different ocean ridge, the short, fragmented Galápagos Ridge, a spur of the much longer East Pacific Rise. Earlier surveys had already spurred fascinating questions focused on this portion of the seafloor. For one thing, the reconstruction of the flip-flopping *magnetic stripes* seemed to indicate that the seafloor was spreading apart much faster in the Pacific than in the Atlantic. As a result, these great cracks in the crust in the Pacific should produce far more heat, and therefore many more eruptions and hot springs, than the massive Mid-Atlantic Ridge.

Moreover, the Galápagos Ridge marked the joint between two small, interesting crustal plates, the Cocos and the Nazca. The piling on of stress had created major fractures and areas where sections of ridges had split and moved independently all along the ridge, raising hopes that the ridge might generate hot springs that would solve the mystery of the missing heat.

The scientists set out to find funding to finance an expedition to this intriguing undersea ridge. They argued that the faster spreading rate should generate hot springs. Moreover, perhaps the superheated, mineral-saturated water would also spur the formation of economically important minerals, in nodules and deposits on the seafloor near the ridges.

Seeking evidence that would support a full-fledged expedition, scientists from the Scrips Institute of Oceanography in 1976 decided to make measurements of the inner *rift valley* of the Galápagos Ridge, a deep cleft running along the top of the ridge. The scientists towed a combination camera and measuring station back and forth over the ridge, hoping to detect a spot where the water was even a fraction of a degree warmer. The thermometer in the towed device did register a handful of just barely warm spots. They also sent back one puzzling photo, a scattering of dead, decaying clamshells that looked like they had been somehow dropped from above. That seemed odd, but the scientists did not pay much attention. The clamshells lay dead and dissolving in the mud more than a mile and a half beneath the surface. No clams could live in such a place, since they must filter their food from the water. Every living thing on the planet gets its energy from the Sun, said biologists confidently. In the ocean, sunlight sustains the plankton, which forms the base of the ocean's food chain just as plants do on the surface.

Scientists pored over the detailed radar maps of the ridge made by the U.S. Navy, hoping to glimpse in the contoured images the most likely spots to find the yearned for hot springs. The radar images and the report of the tiny temperature differences along the ridge eventually prompted the National Science Foundation to come up with several million dollars to fund a full-fledged expedition in 1977.

The scientists turned to a Woods Hole team led by Richard von Herzen and the adventurous Robert Ballard, who had piloted the submersible *Alvin* in the earlier, history-making dive to the Mid-Atlantic Ridge. The team also included geochemist Jack Corliss, who had done much of the pioneering work on the likely effects of the magma-driven chemistry of the deep ocean. Other scientists included chemists Jack Dymond and John Edmond. No one bothered to bring along a biologist to explore the lifeless, volcanic desert of the deep seafloor.

The expedition included the main research ship and two submersibles, the *Knorr* and the *Lulu*, designed to withstand the enormous pressures of water two miles (3.2 km) deep. The big research ship towed the submersibles toward the ridge and then cut them loose to hurry on ahead to lay out a network of beacons on the seafloor near where the earlier expedition had detected those tiny increases in temperature. Those *transponders* were crucial to the upcoming expedition, since they emitted signals that would allow the submersibles to figure out their position more than a mile beneath the surface in the absolute pitch dark. Even after the development of a network of satellites that now allows even a hiker with a handheld GPS to pinpoint his exact location on the surface of the Earth, finding a particular spot on the seafloor posed a huge challenge. How could a craft sink down through two miles of pitch-black water without a single orienting feature and land on precisely the intended position on the seafloor? Any snorkler who has gone up for a breath and then tried to find the same rock on the bottom knows how challenging that can be.

So the mother ship dropped a network of transmitters that would allow the submersible to precisely locate its position and return reliably to a particular spot on the bottom. In addition, it dropped a submarine-measuring device and camera, both attached to a single long cable, and towed it back and forth over the ridge in the vicinity of those telltale temperature *anomalies*. The scientists figured the camera, dubbed *Angus*, would pinpoint interesting locations for a visit by *Alvin*, with its tightly sealed up, badly cramped human crew.

SURPRISE ON THE SEAFLOOR

Angus burbled happily along, beaming its lights at the mysterious bottom and snapping a color picture every 10 seconds to produce a continuous image of the bottom. On the second day of towing, the sensors on the *Angus* recorded a small, three-minute increase in water temperature as

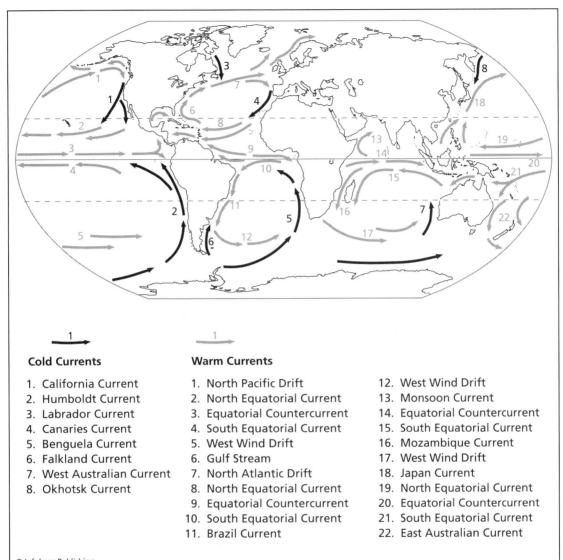

Ocean circulation

the ship on the surface dragged the *Angus* through a patch of warm water. Hoping they had found a hot spring, the scientists eagerly developed the film when the *Angus* returned to the surface at the end of the day, knowing that the transponders and the instruments on the towed camera would allow them to fast-forward to that alluring, three-minute warm spot.

Their wildest imaginations had not prepared them for what they saw. The startling white shapes of hundreds of foot-long clams sprang into focus in the deep, dead, pitch dark of the ocean bottom where no clam should be able to survive. Yet these clams were alive, not dead shells dropped down from above. The picture made no sense at all.

Feverish with excitement, Ballard readied the submersible crews for a descent to that impossible clam bed the next morning. The *Alvin* fell through the darkness for 90 minutes. Thanks to Ballard's experience and the guidance of the carefully planted transponders, they reached the bottom only 900 feet (274 m) from the clam bed *Angus* had filmed the previous day.

Not even the photos of the clams adequately prepared them for what they found, according to Ballard, who recounted the adventure in his book *The Eternal Darkness*. The ocean floor in the glow of the submersible's headlights stretched in a jagged, volcanic jumble, similar to the landscape of the Mid-Atlantic Ridge. The scientists descended to the bottom in *Alvin*, the durable submersible that had also played a leading role in exploring the Mid-Atlantic Ridge (as shown in the color insert on page C-2 [top]). *Alvin* was linked to its surface-based host ship, the *Lulu*, by an acoustic telephone. Peering through the portholes, Ballard could see the cracks and *fissures* in the seafloor shimmering with the heated water of a hot spring. The water from the hot spring was tinted an eerie blue with its load of dissolved *manganese* and other minerals.

Festooned all around that shimmer of superheated water, miles from the faintest glimmer of light, giant, foot-long clams huddled near beds of

---

THE THEORY OF EVOLUTION

The eruptions that created the undersea *vents* that revolutionized theories about the origins of life also created the islands that helped scientists understand how life evolves.

On the *Galápagos Islands*, shown in the color insert on page C-2 (bottom), biologist Charles Darwin made the observations that played a key role in the development of the theory of evolution, the great, organizing principle of modern biology.

In the Galápagos, Darwin discovered strange, isolated creatures that helped him develop his transformative ideas. He found many creatures on the nearby South American mainland, each with sharply different features. That included finches with oddly shaped beaks, gigantic tortoises, and seagoing iguanas as comfortable in the water as alligators.

Darwin suggested that long ago the ancestors of these creatures had become trapped on the islands. Over time, random mutations changed their features. Although most mutations either make no difference or cause problems, some confer a competitive advantage, like webbing between toes or a beak shaped to crack a certain, locally abundant seed. If those mutations give creatures an advantage in producing more offspring, they will pass that trait along to their children and their children's children's children. Eventually, those adaptive changes will spread throughout the population, enabling a species to evolve to meet the challenge of the environment.

long, brown mussels. Little swarms of pale shrimp jetted past the thick view ports. White, ghostly crabs scuttled over the clam bed, picking delicately with oversize claws. A small, pale, lobsterlike creature picked its way over the rocks. Strange fish drifted and darted, flashing into the floodlight then vanishing into the enveloping darkness. Fleshy, fringed, anemone-like creatures covered the rock surfaces near the steaming fissures. Bizarre, fleshy, flowerlike creatures, immediately dubbed "dandelions," swayed on long stalks. Masses of squirmy-looking worms pulsed and writhed.

The astonished Corliss exclaimed, "Isn't the deep ocean supposed to be like a desert?" The scientists had discovered the hot springs needed to validate the theory of plate tectonics, but they had also stumbled over an equally profound and far more startling discovery—a complete ecosystem powered without sunlight. Clearly, they should have brought a biologist.

The astonished and electrified scientists immediately set to work trying to make sense of their baffling findings and to gather samples and information that promised to spawn dramatic new findings on life and its origins. Ballard maneuvered *Alvin* to the bed of clams, delicately using the robot arm of the ungainly looking submersible to grasp first a clam, then a few rocks covered with a matlike, mysterious growth and starfish. He dropped the samples into *Alvin*'s sample bin for eventual return to the laboratories on the host ship (the arm of a submersible is shown collecting starfish in the color insert on page C-3 [top]).

That dive marked the beginning of a heady period of exploration, as scientists discovered an unsuspected world, fueled by the same titanic forces that have continually reshaped and resurfaced the planet. On two trips, the researchers made 24 dives in as many days, eventually exploring five different sites where hot springs nurtured a rich, strange, and utterly unexpected ecosystem.

The water simmering up from those deep cracks in the surface of the Earth was heated to only 63°F (17°C), still a remarkable degree of warmth considering that the surrounding ocean was chilled to just a shiver above freezing and so compressed by the weight of overlying water that it would have instantly crushed any normal submarine.

BEWILDERING CREATURES DISCOVERED

The explorers found a bewildering variety of creatures, with each of the sometimes widely separated sites boasting its own unique characteristics. They found one dying site, littered with slowly dissolving, empty clamshells, evidence that the hot springs both started and died, driven by some change deep beneath the ocean floor. That made sense. This stretch of the longer East Pacific Rise lay at the junction of two plates. The seafloor was

> ### STRANGE REPRODUCTION IN THE DANDELION PATCH
>
> The expedition to the hot springs of the Galápagos Ridge discovered many bizarre new life-forms. One of the strangest was a flowerlike jellyfish they dubbed a "dandelion." The jellyfish dominated one vent they called the "Dandelion Patch." The creatures rose in swaying patches from the seafloor near the heated vents. They turned out to be a new type of jellyfish, related to the deadly Portuguese man-of-war. Biologists later discovered that this strangest of jellyfish lives moored to the bottom by a thread of tissue. The head, ringed by "petals," floats off the bottom because it encloses a small pocket of gas. Each petal serves a different purpose. Some capture microscopic prey, some digest the food, and others carry out the creature's bizarre sex life.

spreading open at eight inches (20 cm) a year along this stretch, one of the fastest spreading rates on the planet. So the constant jostling, shifting, and slipping at these two vast continental plates continually altered the plumbing of the *fissures*.

The scientists dubbed one site the "oyster bed," when the oceanographers, geologists, and geochemists in the group mistook what turned out to be a unique species of mussels for relatives of oysters.

The "Garden of Eden" proved the most impressive and unexpected site. Here they found the dominant clustering of clams, crabs, and dandelions. But they also found a riot of new species, including small pink fish and small *limpets*. They also found clusters of "tube worms," long, tubelike creatures growing out of stalklike shells with a fringe protruding from the end of the tube-shell (shown in the color insert on page C-3 [bottom]). The projecting fringe, waving in the heated currents rising from the vents, was a vivid bloodred. Later, biologists discovered that a specialized form of *hemoglobin* similar to that found in human blood gave the fringes their bright red color. The hemoglobin in the tube worms served the same function as it does for humans, carrying oxygen to fuel *metabolism*. Some of the tube worms were nearly two feet (.6 m) in length. "The animal itself filled more than half this elongated stalk. With no eyes, no mouth or any other obvious organs for ingesting food or secreting waste, and no means of locomotion, it was no worm, snake, or eel, but no plant either, the strangest creature we had ever seen," wrote Ballard in *Eternal Darkness*.

In 1979, a second Woods Hole expedition returned to the Garden of Eden with a supply of biologists. Ballard came along again, this time with a National Geographic Society film crew to work on a TV special, *Dive to the Edge of Creation*.

They had difficulty on this second dive even finding the scattered oasis that had amazed them the first time. They could use satellites to locate their position on the surface of the ocean, but the transponders they had left on the bottom had long ago stopped working. So they had to again use *Angus* to try to locate familiar filmed landmarks from the first expedition. They soon found an even more astonishing site that they quickly dubbed the Rose Garden. Here they found gigantic tube worms 12 feet long (3.7 m), their bloodred fringes ruffled by the rising warm water.

The lava fields of the Galápagos Rift created a similar landscape, but the hot-water oases made it more like an alien planet. Here the scientists discovered living "hedges" clustered around the hot-water vents, composed of a tightly bunched, swaying row of 10-foot- (3-m-) tall tube worms (as shown in the color insert on page C-4 [top]). The heated water, tinted blue or milky white with minerals, rose up in a shimming cloud around them, lending an otherworldly feel to the scene. The eager biologists gathered samples and made measurements, building on the results of the first expedition but with far more expertise and instruments designed to measure living things instead of volcanic rocks.

A WORLD RUN ON SULFUR

The key to the unique ecosystem of the hot-water fissures lies not in the temperature, but in the nutrients. The strange, heat-driven chemistry of the undersea volcanic vents does not depend on sunlight at all. Instead, the energetic chemistry of hydrogen *sulfide* drives the process. The same chemical that makes rotten eggs smell drives the chemistry of what may turn out to be the first ecosystem, perhaps the place were life began.

The expedition's scientists dimly understood that dynamic when they examined the first samples in the ship's laboratory. Everyone in the room reacted immediately to the overpowering stench of rotten eggs. Clearly, the chemicals that sustained the vent creatures came from a volcanic source deep beneath the surface, where seawater under extreme pressure seeped down through the fissures of fractured rock until it came into contact with the magma. As the seawater neared the magma and became superheated under high pressure, the chemistry changed. First, the heated water dissolved many minerals in the surrounding rock, such as manganese. In addition, many elements in the original seawater precipitated out of the water and into the new rock. That dramatically changed the chemical composition of the water that went boiling back up toward the surface after contact with the *magma* supercharged it with energy.

As a result, the stable sulfates found in all seawater were converted to the much more unstable and reactive hydrogen sulfide. This takes place

> ## LIFE RUNS ON SUNLIGHT
>
> Most life on Earth runs on sunlight, even deep in the dark ocean. Creatures that live on the bottom of the ocean must consume the bits and pieces of life-forms that live and die in the sunlit regions and then sink to the bottom. Therefore, sunlight drives the ecosystem of the vast oceans just as it does the rain forests on land. First, plankton converts sunlight to energy. Next, a host of other creatures eats the plankton, ranging from the shrimplike krill to the blue whale. Finally, a host of other life-forms eats the plankton-eaters, especially the krill. A whole food chain depends on the plankton's knack for making energy from sunlight.
>
> The plant chemistry that drives *photosynthesis* allows plants and plankton to use the Sun's energy to break apart molecules of carbon dioxide to produce a host of nutrients. In the process, plants also produce free oxygen as a waste product. The production of oxygen by plants provides the highly reactive, energetic oxygen molecules that sustain the metabolism of most other living things, including human beings.
>
> Early expeditions to the seafloor found few signs of life in the deepest areas because only a faint drift or flux of the sun-spawned nutrients survives the long fall from the upper waters to the deep ocean. Therefore, the vast seafloor is a virtual desert except at the vents, where the food chain runs on sulfur-eating bacteria.

when sulfur binds to hydrogen atoms, a process that occurs only at high temperatures. The resulting brew would instantly broil any normal living creature—but not the tube worms and the other strange forms of life that evolved to suit the conditions at the vents.

The key to the system lies in an unusual form of bacteria that can turn hydrogen sulfides into energy, just as plants use sunlight to turn carbon dioxide into nutrients. The bacteria have evolved into a biological battery powered by a series of chemical reactions that turns the hydrogen sulfide into nutrients, a process called *chemosynthesis*. The reactions also take advantage of the carbon dioxide and oxygen found in seawater.

Scientists had previously discovered similar chemosynthetic bacteria in a few, exotic habitats, such as the hot springs, *geysers*, and pools of Yellowstone National Park. But now they discovered bacteria that occurred in such numbers they could feed an entire ecosystem. All told, on these first two dives to the Galápagos Rift, scientists collected and described 200 different types of bacteria, the ancient and adaptable organisms that made all the other strange creatures possible. Many biologists now believe that a vast population of these bacteria may permeate the fractured rock along the undersea ridges.

Some biologists believe that if it was possible to scrape up all the bacteria hidden away in the seafloor cracks on all of the ridges and trenches, they would weigh more than all the other living creatures on the Earth combined.

The surprised biologists discovered that the warm water roiling up from the fissures had 300 to 500 times more nutrients than the surrounding seawater. Years of research had demonstrated that rift bacteria come in many types, each with a specialized environmental niche. Some live only deep inside the vents, happily drawing energy from water that would boil any other living cell. Other bacteria live on rocks close by the vents, where the temperature is about the same as a cool winter day in Phoenix, Arizona, although the temperature drops dramatically within a foot or two of the vents themselves. A third set of bacteria live inside the bodies of other vent-dwelling creatures, such as the tube worms and clams.

This third type of bacteria maintains a strange, symbiotic relationship with the host creatures. The clams, tube worms, and other creatures offer the bacteria a stable environment, protected from great shifts of temperature and predators. Often the host creatures live only where conditions are perfect for the bacteria they harbor inside their bodies. For instance, the tube worms always position themselves to let the warm,

The strange assemblage of creatures that live alongside hydrothermal vents subsists in one of the only ecosystems on the planet not fueled by sunlight. *(National Oceanic and Atmospheric Administration)*

> ### DID LIFE ORIGINATE ALONG THE VENTS?
>
> Some biologists think that life started at a network of undersea vents more than 4 billion years ago. One of the great mysteries in biology is how life got started so quickly after the planet cooled down. The Earth is about 5 billion years old, but the first fossil evidence of bacteria-like life goes back to perhaps 4 billion years. At that time, the surface of the planet was a violent, dangerous place. Because no living things existed to release oxygen into the air, the Earth did not have a protective layer of ozone to block cosmic rays. Therefore, the surface of the Earth was continually bombarded by lethal particles expelled by the Sun. Cosmic rays, volcanoes, and asteroid impacts made the Earth's surface a hostile environment for the first emergence of life.
>
> But suppose life started instead along this network of undersea vents, protected by the ocean waters and nourished by the chemistry of the vents themselves. Many biologists are now convinced that such vents spawned the first complex life-forms and have nurtured them ever since. So if some inconceivable disaster does strike the surface, the bacteria and tube worms and other bizarre life-forms will remain, safe in the boiling water vents of the deep oceans.

hydrogen sulfide–rich water flow up through the tube and through the cavities in their bodies, where they host colonies of the chemosynthetic bacteria. Meanwhile, the bacteria produce nutrients from chemical reactions that break apart the hydrogen sulfide.

In return, the bacteria release their own "waste" products, which feed the tube worms, clams, and other host creatures. In much the same fashion, plants produce oxygen as a waste product, which makes possible deep breathing by human beings.

Neither the bacteria nor the tube worms could survive without their partner, just like human beings could not survive without a stomach full of bacteria that helps digest the food they eat. The tube worms absorb oxygen and other inorganic compounds from the water, using the red, hemoglobin-rich tips of their body. The red tips look like tiny heads, but they are really more like gills composed of hundreds of thousands of tentacles. The bacteria live inside those tips, eagerly absorbing the oxygen and other vital elements filtered from the seawater by the tube worm.

In that sense, the creatures of the undersea vents do remain at least loosely connected to the sun-driven ecology of the surface. Originally, the Earth's atmosphere had no free oxygen. Once photosynthesizing plants evolved and started releasing free oxygen as a waste product, all the other living things developed to take advantage of the highly reactive oxygen. Only a few types of bacteria managed without oxygen once it became available. The bacteria that form the basis of the food chain in the undersea vents can use free oxygen as well to speed up their *metabolism*. But they do not have to have it.

Repeated dives over the course of two expeditions to the Galápagos Rift gave scientists their first close look at this intricate ecosystem. They found the superheated vents crowded with bacteria, turning sulfur to living tissue and then floating up in the milky, mineral-rich water to sustain all the other rift-dwelling creatures. In the openings of the vents grew clusters of limpets. These caplike relatives of sea snails are found all over the oceans, but the vent limpets looked exactly like fossil relatives biologists assumed had been extinct for millions of years. Close by grew beds of mussels. Near the vents crowded the hedgelike colonies of giant tube worms, anchored to the rock and diverting the sulfur-, mineral-, and nutrient-rich heated water through their long tubes to their symbiotic bacteria.

Farther away from the vent openings grew the beds of giant clams that had given the first, strange hint of the discoveries to come. Most grew along fissures, their back end pointed upward and their opening ends pointed down into the crack so they could filter nutrients from the water rising through the elaborate plumbing of the vents. The biologists marveled at the size of the clams. Even more startling, when they got samples back to the lab and opened them up, their flesh proved as bloodred as the tops of the tube worms. Suffused with oxygen-hoarding hemoglobin, the red, fleshy tissue of the clams enabled them to store the scarce traces of free oxygen and to deal with the potentially toxic effects of the sulfur in the water rising from deep in the Earth.

Some of the clams, mussels, and other creatures also had their own species of enclosed, symbiotic bacteria to help convert the sulfur into usable energy. The clams ultimately provided clues to the development of the vents themselves, since they could survive after the vents shut down, and their slowly dissolving shells indicated places where vents had turned off entirely.

The clamshells eventually proved that the vents along the ocean rifts start and stop continually. Most last for only a few years, perhaps a few decades, before shifts deep beneath the surface shut off the flow of superheated water from where the seawater comes up against the magma.

As a result, the vent creatures evolved fascinating adaptations to the fitful nature of the vents. Many produce tough, sealed, seedlike spores that can drift for long periods on the slow currents of the deep ocean. Many also rise in the heated plume of the vents and then settle over a wide area. Those that end up near another vent can sprout and sustain a cycle that continues, unbroken, for billions of years.

In a particularly bizarre discovery, biologists found that some bacteria closely related to the vent species thrive on the decaying, oil-rich skeletons of whales that have settled to the seafloor. It is possible that key vent species essentially use these scattered whale carcasses to hopscotch across the ocean from one vent field to another. Other vent creatures produce tiny, free-swimming larvae that can cross great distances in search of another plume of warm water, which they can then follow to find a new home.

The dive to the Galápagos Rift completed an undersea revolution that started on the Mid-Atlantic Ridge, spread to the collision of crustal plates at San Juan de Fuca, and finally centered on the exotic creatures living on the heated seafloor close by where Darwin had revolutionized the study of biology.

The 20-year span of undersea discoveries had revolutionized geology, seismology, geochemistry, and now biology. Still stranger discoveries lay ahead, as scientists scattered across the planet to work out the unexpected lessons of this planet-girdling networks of undersea ridges and trenches.

5

East Pacific Rise
Pacific Ocean

The discovery of bizarre forms of life on the vented seafloor just off the coast of South America along the Galápagos Ridge triggered an era of exploration, especially along the massive East Pacific Rise. Minerals venting from the seafloor can be seen in the color insert on page C-4 (bottom). This vast chain of undersea mountains remains the planet's only geological rival to the Mid-Atlantic Ridge, whose towering heights and strange symmetry had spawned the *plate tectonics* revolution.

The East Pacific Rise starts near *Antarctica* and runs in a jagged zigzag arch north past South America, where the Galápagos Rift splits off as a spur that separates two small *crustal plates*. But the East Pacific Rise continues on up to the coast of Mexico, forming the long, enigmatic edge of the largest of all the crustal plates, the Pacific plate.

In the waters of Baja California, the crustal edge that forms the East Pacific Rise changes form. It turns into the long, offsetting fracture of the *San Andreas Fault*, which runs the length of California. The fault plunges back into the sea again to take in the San Juan de Fuca Ridge and runs up the coast of Canada and Alaska. The East Pacific Rise therefore runs all along the eastern edge of the Pacific Ocean, that great void between the continents whose characteristics dominate climate, geology, and the structure of Earth's shifting surface.

The Pacific remains the oldest, deepest, most active, and biologically diverse body of water on the planet. It generates most of the world's earthquakes and *volcano*es and harbors most of its living organisms. You could pour two Atlantics into the Pacific Ocean and hide all the world's continents so thoroughly that only a few Mount Everest–style mountain ranges would stick above the surface. The Pacific accounts for roughly half of the Earth's surface—69.4 million square miles (179.7 million km^2), with an average depth of 14,000 feet (4,270 m). All the other *ocean basins* have a range of mountains running up the middle, such as the mid-Atlantic Ridge. But in the Pacific, the defining crack through

MEASURING GRAVITY

Scientists use complex and sensitive instruments to measure the structure of the Earth below the seafloor. Some of those instruments can detect tiny variations in the Earth's gravitational field caused by the buried edge of a massive crustal plate or great hidden bodies of molten rock in *magma* chambers miles beneath the surface.

For instance, scientists have carefully measured the force of gravity all along the East Pacific Rise, trying to understand the forces that shape the Earth. The force of gravity varies with mass and therefore with the density of the rocks. By measuring tiny variations in gravity, geologists can locate the depth and size of vast chambers of molten rock far beneath the surface, since the semi-molten magma is less dense and therefore has less mass.

*Submersible*s such as *Alvin* could take these *gravitometers* far beneath the surface to measure these changes, but the measurements were so exacting that any movement inside the sub during the hour-long measurement could completely spoil the results.

For instance, U.S. Geological Survey researcher Bill Normark once spent half an hour sitting absolutely still while making one such measurement while the submarine rested on the seafloor. For much of that time, his glasses were sliding, very slowly, down his nose. Finally, he carefully moved to stop them from sliding off and thundering to the floor. Just that small movement ruined the measurement and squandered an expensive dive in the submersible.

The research submersible *Alvin* dangles out of its element, after making repeated major discoveries in its explorations of the deep-sea bottoms. *(National Oceanic and Atmospheric Administration)*

which magma rises and in turn drives the movements of the crustal plates is offset to one side, in the form of the East Pacific Rise. That is because the half an ocean basin to the east of the rise has been largely consumed and overridden by North and South America. That leaves just the eastern half of the ocean basin, slumped down into a great trough as it moves away from the swell in the *crust* of the East Pacific Rise. The great ocean basin stretches more than 8,000 miles (12,875 km) from the shores of Australia and Asia, where the Pacific plate is forced underneath a fragmentation of other plates, leaving a chain of deep-sea trenches up to seven miles (11 km) deep.

Intriguingly, this oldest of ocean basins is also the most active. Measurements of the seafloor *magnetic stripes* along the East Pacific Rise show that the seafloor is spreading apart faster than any place else on the planet. Calculations suggest movement rates of up to eight inches a year, three or four times faster than most sections of the Mid-Atlantic Ridge. That same calculation originally led explorers to the Galápagos Rift; now it draws them to other, more remote areas of the East Pacific Rise.

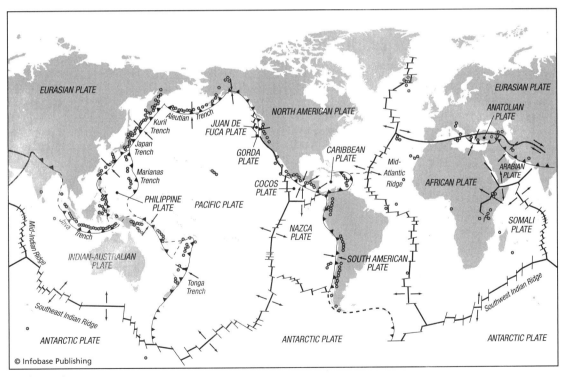

Earth's active zones

Once again, Ballard and a large, diverse team of scientists set their sights on a bold, expensive expedition that would use *Alvin* to sample portions of the rift, confident that the rapid rate of *seafloor spreading* would generate more surprises and more of the undersea *vents* that had so excited biologists along the Galápagos Rift.

RACE TO EXPLORE A BIZARRE WORLD

That feverish search for new vents lured many scientists who had previously been conducting much more tedious efforts to chart the overall structure of the East Pacific Rise. Many scientists jumped at the chance to start looking for more undersea *vents* along the East Pacific Rise during several expeditions between 1977 and 1979.

Previously, research teams prepared for such dives by conducting detailed surveys of these undersea mountain ranges with cameras towed back and forth across the crest of the mountain range. That worked best on the Mid-Atlantic Ridge, but the East Pacific Ridge rose much more sharply from the bottom, due to the faster rate of spreading. Therefore, the camera towed within visual range of the bottom would smash constantly into jagged outcrops and ridges of volcanic rock. The scientists had nicknamed *Angus*, the towed camera, "dope on a rope," partially for its ability to take a pounding. But they did not want to overtax the durable equipment, on which crewmembers painted the motto "takes a licking and keeps on ticking." Besides, biologists and geophysicists agreed that most of the vents would occur in or near the narrow, steep-walled *rift valley* that ran the length of the East Pacific Rise. That mountaintop rift valley averaged about a quarter to half a mile wide. The cameras, gravity, and magnetic and temperature measurements identified three places likely to yield more vents, which biologists hoped would produce more Earth-shaking surprises. A group of French scientists teamed up with researchers from Scripps Institute of Oceanography in California to explore the East Pacific Rise, which was in Scripps's backyard.

The early surveys revealed new surprises in one of the most active, unstable, and high-energy volcanic areas on the planet. The cameras revealed large *lava* flows dominated by the smooth, bulbous *pillow lavas* common to the Mid-Atlantic Ridge and places such as Hawaii, where lava oozes to the surface and hardens into rounded "pillows." The semi-molten lava took distinctive forms, like toothpaste squeezed out of a tube. Researchers also discovered large areas covered by lava that had chilled to brittle, glass-like forms due to contact with the frigid, high-pressure seawater. Again, such formations were common along the slower-spreading, less active Mid-Atlantic Ridge as well.

An early expedition along a portion of the East Pacific Rise revealed another mystery. The French submersible *Cyana*, manned by researchers from Mexico, came upon some crumpled chimneylike structures,

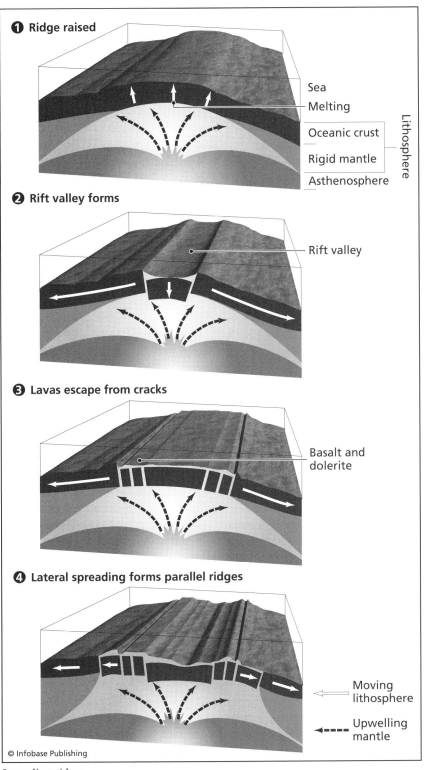

Spreading ridge

THE MAKING OF GOLD AND SILVER

The discovery of mineral-spewing *black smoker*s on the East Pacific Rise soon led geologists to a new understanding of how many minerals are created, including gold and silver. Clearly, the combination of magma and superheated seawater in cracks under the seafloor helped account for the formation of a host of minerals. For instance, galvanized by the discovery of the chemistry of the smokers, geologists took a new look at mineral deposits all over the world. Geochemists Rachel Hayman and Randy Koski studied the 95-million-year-old rocks in one copper mine in Saudi Arabia. They soon discovered that those rocks had been originally forged along a mid-ocean ridge on the bottom of some long-vanished ocean. As proof, they found in the copper-bearing rocks the fossils of tube worms. The discovery excited a rush of interest by people interested in mining the now revealed mineral deposits along the ridges.

One expedition discovered a mineral-rich mound 100 feet (30.5 m) thick, 600 feet (183 m) wide, and 3,000 feet (914 m) long that was topped by two-dozen fuming black smokers. The minerals in that mound weighed millions of tons, with the copper alone valued at $2 billion.

glittering in the spotlight. The researchers collected samples of the enigmatic formation and later discovered the rocks were almost pure sphalerite, made mostly from zinc *sulfide*. The *crystallized* rocks also included 50 percent zinc, 10 percent iron, and 1 percent copper, with traces of gold and silver. The finding offered the first intriguing clue that the superhot chemistry of the undersea vents might produce valuable minerals. No one knew what to make of the discovery at the time, but it would eventually reshape geology and geochemistry.

The discovery posed a geological puzzle. Only temperatures between 600°F or 1,000°F (316°C or 538°C) could precipitate such minerals out of rocks and seawater, nearly 10 times hotter than any vent temperatures ever recorded. The geologists assumed the minerals had been produced far beneath the surface and then somehow pushed up through the vents. The only other explanation was that raw, superheated magma had erupted right at the surface, an idea that seemed wildly implausible and in conflict with all of the undersea expeditions so far. All of that set the stage for the dramatic expeditions to the East Pacific Rise that would once again demonstrate the unsettling extent of ignorance about these remote areas.

MOUNTING A HISTORIC EXPEDITION

Scripps and a group of French scientists joined forces for the historic exploration of the rift. The Scripps Institute expedition to a potential vent site included *Alvin* pilot Dudley Foster and French volcanologist Thierry Juteau. They found deposits of sulfides that offered evidence of extinct vents, but no signs of life.

On their way to the second site on a subsequent dive, they noticed thickening scatters of ghostly white crabs, scuttling across the otherwise barren seafloor. As they approached the location of a temperature spike that indicated the presence of a vent at a depth of 9,000 feet (2,740 m), they noted a decrease in visibility. However, they saw no sign of the rich assemblage of living things they had found at the Galápagos Rift. Suddenly, a strange object drifted into view through the thick view ports. Rising from the seafloor was a narrow, six-foot- (1.8-m-) tall smokestack that billowed dense black smoke. Foster described it as "a locomotive blasting out" smoke. They realized that this could be an active version of the dead formation the earlier expedition by the *Cyana* had discovered. The thermometer at the end of *Alvin*'s mechanical arm immediately registered a noticeable rise in temperature.

The scientists all stared in baffled wonderment at the chimney, later dubbed a "black smoker" (shown in the color insert on page C-5 [top]). No one had predicted such a feature. Everywhere else scientists had gone, heated water simmered up from cracks in the rock. Here it seemed that the hot water, laden with opaque minerals, had generated a tube, as the minerals in the water percipitated out of solution when the vent water escaped into the ocean.

Cautiously, Foster maneuvered closer to the black smoker, hoping to measure the temperature of the water and perhaps take a sample for later analysis. *Alvin* had a series of pressurized ballast tanks so that it could come to a standstill and float motionless above a selected spot on the seafloor. This made it highly maneuverable but also left it vulnerable to any sea bottom current. As *Alvin* approached the smoker, two things happened in quick succession. First, the outside thermometer failed. Second, Foster found himself fighting a strong current that was sucking the submersible toward the black smoker. Later he realized that the updraft of warm water from the top of the six-foot- (1.8-m-) high chimney of crystallized stone was causing cold, heavy water to rush in toward the smoker, creating the current that had *Alvin* in its grip.

Acting on instinct, Foster decided to go with the flow and let the inrush to the updraft pull *Alvin* into the dense smoke, which would then cause the submersible to rise on the current. He did not worry about the temperature of the water, since none of the vents on the Galápagos Rift had measured hotter than a mild 73°F (23°C).

Immediately, all landmarks vanished in the billowing clouds of black smoke. Foster struggled to regain control, and for a moment, the sub was trapped directly over the furious spew of the vent, held in place by the inrush of colder water from all sides. As Foster maneuvered, *Alvin* smashed against the top of the chimney, which shattered and toppled with the impact. Foster took on more ballast, and *Alvin* settled to the bottom near the toppled chimney, coming to rest on a gently sloped, 30-foot- (9-m-) wide mound of rock composed of hundreds of other toppled

Scientists load gear into *Alvin* before a dive to the bottom of the sea. *(National Oceanic and Atmospheric Administration)*

> ## LAKES OF LAVA
>
> Explorations of the East Pacific Rise revealed new landscapes, including the enigmatic remains of an undersea lake of lava. Initially, geologists were baffled by the discovery of large, hollowed-out, steep-walled depressions, with mushroom-shaped pillars scattered throughout the inside of the depressions.
>
> Later, geologists figured out the origins of these otherworldly features. Clearly, molten rock had erupted from magma chambers that had pushed much closer to the surface here than in the Atlantic. Masses of this superheated volcanic rock formed incandescent lakes of lava. Contact with the overlying water cooled a rocky crust atop the lakes of lava, allowing the magma to remain molten for some period. Once the eruption that had forced the magma to the surface subsided, the still-molten rock drained into the subsiding magma chamber, leaving an empty chamber where the lava lake had been. The strange pillars in the middle of the lake were places where the rock had hardened as the lava lake cooled and then retreated. Once the molten rock drained back down into the Earth, the roof of the lava lake collapsed, leaving the depressions with the mushroom pillars that the researchers had photographed. The cameras even recorded odd bathtub rings around the sides of the depressions that marked different shorelines of the molten rock, which may have repeatedly advanced and retreated into the lava lakes from the network of *fissures* connecting them to the magma chambers below. Mount Erebus, concealing one of the few standing lava-filled lakes in the world, can be seen in the color insert on page C-5 (bottom).

structures. The rock glittered and gleamed in the headlight, suggesting that the chimneys were made of minerals that had crystallized out of the mineral-saturated, superheated water of the vent, just as stalactites in caves are made of minerals that condense out of saturated, pressured water that drips into a cave. Foster delicately collected several samples. Then he turned his attention back to the vent and cautiously advanced the thermometer into the vent water. The readings quickly rose to the 91°F (33°C) maximum of the thermometer, nearly 20° hotter than the maximum 73°F (23°C) recording in the Galápagos vents.

The high temperature worried Foster, who peered out through the view ports that held back the thousands of pounds per square inch pressure of the overlying two-mile- (3.2-km-) high column of water. He conferred briefly with engineer Jim Akens, up on the surface. Akens convinced Foster that the thermometer must have broken. Reassured, Foster and the other scientists set off to find another vent. This one proved much like the vents of the Galápagos Rift, well behaved and lacking in the dramatic black smoke.

NASTY SURPRISE AWAITS

When they returned to the support ship, the crew of the *Alvin* discovered that they had narrowly cheated death. When Akens pulled out the malfunctioning thermometer, he was astonished to find it half melted and dramatically charred. Alarmed, he inspected the underside of the *Alvin* where it had floated directly above the black smoker. The fiberglass on the outside of the hull had started to melt. Baffled and alarmed, the engineers thumbed furiously through the manuals and discovered that the vent water had to be at least 356°F (180°C) to melt the polyvinyl chloride plastic.

In short, if that superheated vent fluid had blasted against *Alvin*'s Plexiglas view point, it might have melted or at least weakened enough to be shattered by the inexorable pressure of the water bearing down on the sub. The news proved the opening chapter of yet another round of revolution and astonishment, delivered by the always-surprising explorations of the deep sea. Repeated dives revealed a network of these black smokers, which one amazed scientist declared seemed "connected to hell itself." Dives also revealed "white smokers" that emitted white smoke based mostly on temperature and mineral content differences. The scientists scrambled to get tougher thermometers and soon discovered a maximum water temperature of 662°F (350°C), nearly 10 times as hot as the vents on the Galápagos Rift—and hot enough to melt lead or the windows of the submersible.

The discovery of the superheated black smokers finally provided the missing heat energy the geophysicists needed to put the final touches on the theory of seafloor spreading, which could account for the movement of continents and the raise of 40,000-mile- (64,370-km-) long mountain ridges. The scientists also soon realized that the boiling chemistry of these vents affected the composition of seawater and deposited many

WHERE DOES THE OCEAN GET ITS SALT?

The discovery of the black smokers on the East Pacific Rise solved the vexing mystery of where the ocean gets its salt and other trace elements. For decades, scientists had gathered masses of information about Earth's water cycle. In that cycle, evaporation distilled pure seawater, which then fell as rain. On its trip back to the ocean, the water dissolved various salts and minerals from the land and so returned them to the sea.

For a long time, that made perfect sense. But then scientists studied the composition of the rivers more carefully and concluded they could not account for the salt and minerals in seawater. Certain chemicals and minerals found routinely in seawater did not occur at all in rivers. Even elements that did occur in river water seemed far more concentrated in the ocean. For sometime, the chemical composition of seawater remained a frustrating mystery.

The smoking undersea vents demonstrated that all the missing elements in seawater had billowed and fumed out of the smokers, including pyrite, anhydrite, chalcopyrite, sulfur, zinc, iron, copper, lead, gold, silver, and other metals. Clearly, such furiously spewing vents on ocean ridges all over the planet controlled the chemistry of seawater.

valuable minerals when the minerals precipitated. This extravagant brew of minerals accounted for the rapid growth of the chimneys of the smokers. Researchers recorded one vent that grew 16 feet (5 m) of chimney in a single year. Some vents built "pagodas," bizarre structures 30 feet (9 m) high with multiple, smoking vents.

Strange new creatures inhabited this hellish landscape, spewing precious metals from the very heart of the Earth. Pompeii worms lived in massive, honeycombed tubes clustered by the vent. The worms moved freely through the maze of tubes attached to the walls of the hottest chimneys, subsisting in the most extreme environment.

Scientists began to calculate the implications of mineral-saturated water hot enough to melt lead venting directly into the deep sea. Once scientists worked out the math and estimated that all the water in the ocean circulates through this boiling network at least once every 8–10 million years. In the process, the superheated water dissolves and redistributes from the miles of buried rock all the vital elements necessary to sustain life on the entire planet, including iron, copper, zinc, and *manganese*. Every cell in the body contains trace elements of these minerals, each essential to the proper functioning of the cells themselves.

Suddenly, geologists realized that the constant recycling of these minerals through the undersea vents accounts for their abundance everywhere on the planet. The rush of money from investors hoping to gather up a fortune in minerals from the sea bottom provided funding for additional expeditions, including a comprehensive survey of the East Pacific Rise by Ballard and several colleagues.

They began with the assumption that the areas with the fastest rates of spreading would generate the most vents and minerals. They pondered the layout of the rift system that the originators of plate tectonics had demonstrated with handmade paper models years earlier. The long central fracture of the rift that formed the edge of the Pacific plate was constantly crosscut and offset by other long faults, running at near-right angles to the generally north-south trending line of the East Pacific Rise. So the central rift valley proceeded in an endless set of jagged zigzags, each interrupted by a fresh set of deep east-west running faults and canyons. The central rift puckered upward between these offsetting faults, forced up to within a mile or two of the storm-tossed surface of the Pacific. Ballard and his colleagues reasoned that this rising blister of rock demonstrated that the superheated plume of magma must come closer to the surface in this stretch, which should yield a greater density of vents and minerals.

Their theory proved accurate and so finally filled in the last puzzle piece of plate tectonics by revealing the location of the missing energy needed to remake the planet. Geologists now estimate that every year undersea volcanoes mostly along these great rifts and trenches spew out enough lava to cover an area four times greater than Alaska to a depth of three feet.

Arctic Ridge

Arctic Ocean

Finally, geophysicists thought they understood undersea trenches and ridges and the vast *convection currents* in the semi-molten rock of the Earth's *mantle*. They confidently predicted that the speed of *seafloor spreading* would govern seafloor volcanic activity. And thus attention turned to the slowest-spreading ridge on the planet: the Gakkel-Nansen Ridge that runs for 1,100 miles (1,770 km) from Greenland to Siberia beneath the ice-locked Arctic Ocean.

The Arctic Ocean has a complex ocean floor geography dominated by three great, parallel ridges, each running roughly north-south for distances of about 1,000 miles (1,609 km), including the Gakkel-Nansen, the Lomonosov Ridge, and the Mendeleyev Ridge. Each long ridge is flanked by a deep, flat basin. These basins average about 12,000 feet (3,660 m) deep. Each ridge is topped by the familiar *rift valley* of a spreading ridge, some of them only about 600 feet wide (183 m) at the tip.

The long, steep Gakkel Ridge dominates the complex topography of the bottom of the Arctic Ocean and remains the slowest spreading of all the major undersea ridges. The *crust* along that ridge spreads

---

THE ARCTIC OCEAN

The smallest of Earth's oceans, the ice-sheathed Arctic is about the size of China, an icy cap to the top of the world that covers the North Pole. In fact, the magnetic North Pole lies on the seafloor 13,000 feet (3,960 m) beneath the surface of the ice (see the color insert on page C-6 [top]). Locked into place between Siberia, Alaska, Greenland, and Scandinavia, the Arctic Ocean until recently remained with three to 10 feet (1 to 3 m) of ice year-round. It nurtures a rich ecosystem that includes seals, polar bears, walrus, whales, and rich feeding grounds of tiny, shrimplike krill. The extreme conditions of the North Pole can be seen in the color insert on page C-6 (bottom). Despite its small size, the Arctic plays a crucial role in regulating the climate of the whole planet and the food chains that sustain life throughout the oceans.

74 ✦ Ocean Ridges and Trenches

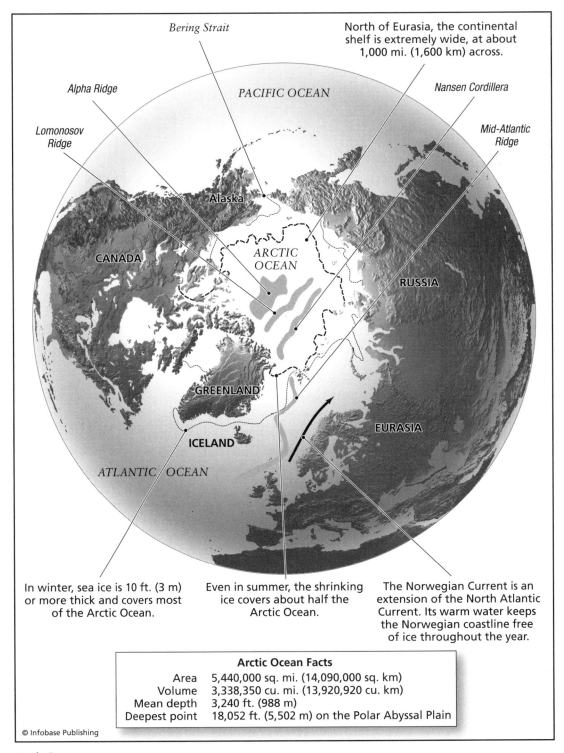

Arctic Ocean

apart at less than half an inch (2.5 cm) a year, compared to some eight inches (20.3 cm) a year on the fast-spreading East Pacific Rise. That slow-spreading rate attracted geophysicists anxious to put the finishing touches on theories of *plate tectonics*.

By the end of the century, the detailed theory of plate tectonics envisioned great, circulating currents in the semi-molten mantle, coupled in some way to the vigorous heating of rock in the Earth's fully molten *core*. Currents in the liquid core drove *convection cells* in the mantle, which created the network of ridges and trenches when the hot, rising *magma* encountered the solid rock of the crust. Geologists believed these slow, irresistible circular currents pressed up against the brittle crust where the ridges formed along zones of weakness. Those currents heated and swelled the solid rock of the crust, forming the great bulge along those 55,000 miles (88,500 km) of ridges. Where the mantle currents came closest to the surface, magma sometimes broke through to the surface, building up the central features of the ridges. Where the convection currents turned and headed back down, they drew crust with them, creating the trenches.

Geophysicists saw a wonderful opportunity to test the theory along the Gakkel Ridge some 3,000 feet (914 m) beneath the surface of the Arctic ice sheet near the North Pole. They also saw a chance to get hold of a piece of the Earth's mantle, to solve the layered mystery of the planet's evolution. The semi-molten rock of the mantle is heavier, denser, and composed of different minerals from the light continental crust rock that floats on the surface of the mantle. Bits and pieces of the mantle rock that sometimes break the surface undergo dramatic chemical changes as a result of the violence of the upheavals that bring them to the surface. However, geologists theorized that the slow, sedate upwelling of mantle rock into the great crack of the Gakkel Ridge would not completely transform the mantle rock. Therefore, if they could haul chunks of rock up from the top of the ridge, they might have relatively pristine samples of the Earth's mantle, which makes up the bulk of the planet.

The theory represented the latest increase in appreciation for what had long been considered a frozen wasteland. In the early days of European exploration, many explorers had hoped to find an ice-free passage through the Arctic Ocean that would open a short route from Europe to North America and on to Asia. But all those early efforts ended in ruin at the edges of the vast ice cap in a howling, frozen wilderness. The Arctic remains in bitter darkness for half the year, due to the 23 degree tilt of the Earth on its *axis*. That tilt accounts for the succession of seasons in the course of the Earth's annual journey around the Sun.

> ### A WANDERING POLE
>
> One of the goals of early explorers was to reach the North Pole. A succession of later explorers made heroic, often futile, and sometimes fatal efforts to reach the North Pole, which is actually on the seafloor and which shifts position slightly from one year to the next. If one draws a line through the center of the Earth precisely along the axis of its spin, the line would break the surface at the North and South Poles. Geographers and navigators call this direction True North. But because the Earth wobbles a bit on its axis, the precise location of the North Pole wanders in an area about 65 feet (20 m) across, occupying a slightly different position every year.
>
> The magnetic pole is determined by the Earth's much more erratic *magnetic field* and can be 1,000 miles (1,609 km) offset from the North Pole, as defined by the Earth's spin. At unpredictable intervals of perhaps hundreds of thousands of years, the magnetic pole actually flips its orientation, with north becoming south and vice versa.

DARING EXPLORER BRAVES THE ICE

One of the most fascinating of those early characters whose name was attached to the most important of the seafloor ridges was Fridtjof Nansen, a tough, idealistic, ambitious, adventurous man. Born near Oslo in Norway in 1861, Nansen was captivated by the solitude, beauty, and challenge of the Arctic when he first worked on a seal-hunting expedition in 1882. He earned a Ph.D. in biology and dedicated himself to exploring this harsh, otherworldly landscape. In 1888, he and five companions crossed the frozen interior of Greenland, the world's largest island. They used skis, although all the experts insisted it was impossible without a base camp to fall back on and without teams of sled dogs to haul supplies. They endured hardship during their two-month ordeal but were the first explorers to prove the entire interior of Greenland was encased in ice.

Next, Nansen resolved to mount an even more dangerous expedition—to deliberately maroon himself in the Arctic ice sheet. Years earlier, an American ship had been trapped in the ice. Its wreckage floated with the unexpectedly mobile ice from the New Siberian Islands all the way to the southern tip of Greenland, suggesting for the first time that perhaps the North Pole was not a frozen land covered with ice but a single, great ice flow, with only scattered, ice-covered islands of solid ground. Nansen realized that a westerly ocean current must flow out of the Arctic Ocean, carrying ice along with it. He decided to build a ship with a hull strong enough to survive getting stuck in the ice, and then let it drift from Siberia to Greenland, hopefully passing over the unvisited and almost unimagined North Pole.

Nansen constructed the *Fram*, with a stout, rounded hull so that freezing ice would lift the ship instead of crushing it. In 1893, he marooned himself in the ice near Siberia and spent the next year making painstaking scientific measurements as the ship drifted southward with the ice. When his calculations revealed that the ship would pass hundreds of miles from the actual North Pole, he and a companion packed skis and dog-drawn sleds and set off with 100 days of food for the North Pole, leaving the rest of the crew to continue drifting south on the ship.

They struggled for two months across jagged ridges of ice, coping with terrible conditions as the ice continued to drift south. Finally realizing they would die if they continued, they headed again across the featureless ice. They reached solid land some 130 days after they had set out, having come closer to the North Pole than any other known human being. They spent another endless winter there, living in rock shelters and dressing in the hides of the seals and polar bears they killed to survive. When summer again made travel possible, they continued south, almost miraculously encountering a British expedition that took them back to Norway. There, Nansen discovered his ship had indeed floated safely to Greenland in a remarkable three-year journey.

Despite the efforts of a handful of early explorers, the Arctic Ocean remained barely explored for most of the next century—adventurers awaiting technology that could cope with the terrible conditions, the thick ice, and the arduous access.

Scientists began to lift the icy veil that hid the bottom of the Arctic Ocean in the 1990s through surveys undertaken by U.S. nuclear submarines that could cruise for months beneath the Arctic's thick covering of ice. The U.S. Navy launched a five-year survey effort in the 1990s dubbed Science Ice Exercise (SICEX), in cooperation with the U.S. National

EXPLORER AND HUMANITARIAN

Fridtjof Nansen's remarkable exploration of Greenland and his three-year struggle to reach the North Pole while marooned in the ice made him an international hero. But he was as much a hero of science and humanity as an adventurer. As a scientist, he published a massive, six-volume scientific study of the journey, which made major contributions to the study of the Earth's magnetic field, meteorology, oceanography, and zoology. Then he got deeply involved in politics, especially in keeping Norway neutral during World War I. After the war, he played a leading role in helping war refugees and released prisoners of war. Nansen helped secure the safe return of 450,000 prisoners of war to their homelands, worked for the establishment of the League of Nations, and led international famine relief efforts that saved hundreds of thousands of lives. That work eventually won him the Nobel Peace Prize.

Science Foundation. Scientists attached pods to the hulls of the massive submarines that could bounce sound waves off the bottom and then made a map from the echoes off the seafloor, 13,000 feet (3,960 m) below the ice. The Seafloor Characterization and Mapping Pods (SCAMP) provided images 100 times more detailed than any previous soundings. The survey produced immediate surprises.

ARCTIC SURVEY PRODUCES SURPRISES

For instance, the survey revealed deep gouges in the tops of undersea mountains and shelves in relatively shallow water. The deep groves suggested that the Arctic Ocean, which is now covered mostly with a 10-foot- (3-m-) thick ice cap, once had layers of ice thousands of feet thick. Perhaps the ice even rivaled the two-mile- (3.2 km-) thick layer that entombs the South Pole, with thousands of years of snow and ice accumulation so heavy that the continent that sits astride the South Pole has sunk thousands of feet under the weight of overlying ice. The discovery stunned climate experts, whose theories and models did not envision such a mass of ice in the Arctic Ocean even during the deep freeze of the *ice ages*.

Moreover, the survey revealed huge, flat plains extending for up to 750 miles (1,200 km) from the shores of the continents that ring the Arctic Ocean. Most of the continents have a relatively narrow shelf of shallow water extending out from the dry land. These continental shelves often become exposed and flattened by waves and sediment dumped by rivers during ice ages, when the ice caps expand and lock up so much water that sea level falls by hundreds of feet. The unusually wide continental margins in the Arctic Ocean have helped lock the ice cap into place. However, if the ice cap thins, large stretches of now locked-in-place ice could suddenly float free and perhaps break up. That is a worrisome finding given the current warming of the planet and the retreat of *glaciers* and ice caps already recorded worldwide.

The submarine surveys also revealed an unexpected link between the topography of the sea bottom and the qualities of the water in the great currents flowing between the bottom of the ice and the seafloor. For the first time, scientists could measure significant changes in the temperature and salt content of the ocean linked directly to the location of the submerged mountain ranges, valleys, and troughs beneath the surface. Apparently, the ridges of the mountain ranges control currents, create vast pools of seawater with differing salinities, and channel the flow of water in and out of the enclosed space of the Arctic Ocean. That means the great ridges and troughs on the seafloor of the Arctic Ocean may directly affect the planet's climate.

Those currents flowing out of the Arctic Ocean also affect the productivity of the world's oceans. The concentration of nutrients in the Arctic after its dramatic, six-month-long siege of dark winter fosters great blooms of plankton and krill that draw whales from every ocean to feast on the fringes of the Arctic ice sheet.

The Seafloor Characterization and Mapping Pods (SCAMP) also revealed another major surprise, an underwater *volcano* in the act of erupting. First, the *sonar* maps of the Gakkel-Nansen Ridge revealed two cone-shaped mountains mostly free of the thick coating of sediment that covered the seafloor all around them. This thick layering of sediment in the troughs between the ridges provided the best evidence of the sluggish rate of spreading along the ridges. But the lack of sediment indicated these two mountains were twin volcanoes that had sprouted relatively recently from the seafloor. Then a network of seismographs (earthquake detectors) measured a swarm of perhaps 250 small earthquakes centered on those two mounds in the course of 1999.

Scientists had many reasons to be excited in 2001 when the National Science Foundation teamed up with the U.S. Coast Guard and an international lineup of scientists to finally take samples from the seafloor of the Gakkel-Nansen Ridge system. They hoped to answer a host of questions from the slow-spreading ridge system and find samples of the Earth's mantle, oozing out from a landscape they assumed would be cold, barren, and lacking in the undersea *vents* that had nurtured such a riot of life on the East Pacific Rise and the faster-spreading portions of the Mid-Atlantic Ridge.

Once again, the ridges confounded the predictions. Dramatic advances in technology had finally made such a survey possible. One hundred years ago, Nansen could only maroon his sturdy ship and drift with the ice for three years. But the 2001 expedition had use of the U.S. Coast Guard icebreaker *Healy*, named after a former slave who became one of the most famous Coast Guard captains and navigators, renowned for navigating the bewildering, ice-choked passage in the *Bering Sea* and Alaskan Arctic.

The *Healy* was a cold-weather oceanographer's dream. Tough-hulled and 420 feet (128 m) long, the exquisitely engineered bow could slide easily through ice 4.5 feet (1.4 m) thick. By backing up and charging forward, the icebreaker could bully its way through ice layers eight feet (2.4 m) thick. *Healy* boasted a system to automatically lubricate the hull to help it slip through the ice, an anti-roll stabilization tank, and bow thrusters to help it maneuver and crack through the ice. Computers constantly processing signals from satellites overhead and connected to the propulsion systems at both ends of the ship enabled the *Healy* to maintain its position precisely so it could coordinate and pinpoint the continuous sonar mapping of the bottom. In addition, the ship could quickly cut holes in the ice and drop dredges and probes to bring back samples from the bottom.

On board, the *Healy* had a crew of 75 and accommodations for up to 50 scientists, who had the run of 50,000 square feet (4,645 m²) of science labs. The *Healy* also boasted two helicopters and five boats for side trips.

Still, many researchers remained skeptical. The *Healy* was supposed to make continuous measurements of the bottom while proceeding along at three miles (4.8 km) an hour through four-foot- (1.2-m-) thick sheets of ice. Some worried that the din of crashing through the ice would overwhelm the whisper of the sound echoes off the seafloor. And even if the soundings worked, many scientists figured the *Healy* would bring back only dead rocks from an undersea wasteland, devoid of the vents and strange life-forms that had electrified scientists on other ridge systems.

VOLCANOES CHALLENGE THEORIES

Just north of Greenland, *Healy* scientists took their first sample. They dropped the dredge, dragged it along the seafloor, pulled the samples to the surface, and stared at the rocks in surprise—they had found fresh *lava*. They soon discovered that large, active volcanoes dotted the seafloor along the ridge close to Greenland, producing far more volcanic activity than anyone expected based on the measured spreading rate along the ridge.

The surprises piled up as the expedition ambled along the ridge system, stopping to take 200 samples, about once every three miles (4.8 km).

The equipment easily punched holes in the ice, and the noise of the icebreaker did not impede the sonar imaging. Things went so smoothly that in two months they easily completed two passes along the ridge, using the results from the first run to concentrate their attention on the most interesting and active areas on the trip back.

To their astonishment, they detected the warm-water plumes from at least nine to 12 separate undersea vents, twice what even the most cheerful of optimists had predicted. They used a variety of indirect techniques to search for a vent in the frozen ridge. First, they towed a thermometer that could measure tiny changes in water temperature. Next, they attached a light sensor to the sampling dredges that could measure even the slightest cloudiness in the pitch-black water. This sensor could detect the minerals and bacteria that cloud the water near vents. Once the sensors detected a faint increase in cloudiness, the ship would turn and follow the readings along this plume of cloudy water back toward the suspected vent.

The dredges brought up samples of rocks like the delicate, *crystallized*, mineral-rich walls of the *black smokers* found in the Pacific. They also brought up samples of vent-dwelling creatures. The samples provided a tantalizing glimpse of a whole new vent community, since the Arctic Rift system is almost completely separated from the rest of the planet-circling rift systems. Biologists believe the creatures that live in its vents may have evolved into strange new forms or perhaps retained characteristics of the first vent creatures according to Linda Kuhnz, a biologist from Moss Landing Marine Labs in California. Biologists are still studying those samples, trying to determine how closely related the vent creatures of the isolated Arctic are to the planet-wide spread of their cousins along other rifts.

Geologists on the expedition including chemical oceanographer Hedy Edmonds of the University Texas at Austin and Charles Langmuir of the Lamont-Doherty Earth Observatory at Columbia University expressed surprise at the volcanic activity in an area where plate spreading was so slow. Moreover, they discovered unusual tall conical undersea volcanoes from which magma oozed steadily.

The findings forced yet another scramble to update the theory of plate tectonics, which now promised to unlock surprising secrets about the structure of the Earth itself, including the semi-molten mantle on which the lighter rocks of the continents float.

First, they realized that vents, volcanic activity, and the characteristics of the seafloor are affected by many things besides a simple rate of spreading along the ridge. For instance, scientists are still baffled by the discovery that the hard, brittle crust on the Arctic seafloor is much thinner than along other undersea ridges. Sound waves change speed depending on the density of the rock through which they pass as do the waves of energy set off by earthquakes and explosions. That has allowed geologists to find the boundary between the brittle, less dense rocks of the crust and the plastic, more dense rocks of the mantle. In addition, measurements of tiny changes in gravity based on the mass of the underlying rock also help pinpoint where the crust gives way to mantle. Scientists have now concluded that while the crust is about 3.7 miles (6 km) thick along most of the ridge system, under the Arctic the crust is only 1.2 miles (2 km) thick. It might thin to almost nothing in places. In effect, along parts of the Arctic Ridge the mantle that remains deep beneath the cool, brittle, chemically different rocks of the light crust actually breaks through to the surface, cooled by its rise toward the surface but little changed chemically.

"What we found could not have been extrapolated from decades of previous studies of the ocean ridge system," said chief scientist Peter Michael, from the University of Tulsa. "It still shows there's much to

> ### HOW UNDERSEA RIDGES CAN AFFECT THE CLIMATE OF A PLANET
>
> Recent studies suggest the Arctic Ocean may play a much stronger role in regulating the Earth's climate and the ocean's productivity than scientists previously suspected. Many climate experts now believe that the geology of the Arctic Ocean helps drive vital ocean currents that reach all the way to the equator. That is largely an accident of topography. The Arctic Ocean is almost landlocked, which is why it hides under an almost continuous ice cap anchored to the surrounding continents. A deep, broad opening between Greenland and Scandinavia connects the Arctic Ocean to the Atlantic Ocean, providing an opening for cold, nutrient-rich deepwater currents. That opening, in turn, is maintained largely by the seafloor spreading along the northern extension of the Mid-Atlantic Ridge north of Iceland that fragments into the complicated system of slow-spreading ridges and troughs on the floor of the Arctic Ocean.
>
> However, the seafloor ridges block much of the deep, cold water from the Arctic from flowing into the Atlantic. This creates a vast, rare pool of super-chilled water trapped in the bottom of the Arctic Ocean. Meanwhile, warmer surface waters move back and forth freely from the Arctic to the Atlantic, creating crucial, fast-moving ocean currents. This drives the chilled, East Greenland Current and weaker secondary currents. The chilled water flowing out of the Arctic also has less salt and minerals than normal seawater, which makes this water lighter so it rides on top of other water in the Atlantic.
>
> The arctic waters leave much of their salt and minerals behind in the super-chilled water that pools on the bottom of the ridge-walled ocean basin. As a result, the cold surface East Greenland Current flows out of the Arctic and chills the northwest shores of Europe and North America. In the meantime, other warmer currents driven in part by the pressure of these arctic currents effectively warm up many areas in Asia and Europe. The undersea ridges of the Arctic Ocean therefore play an important role in the climate of the entire planet.

be discovered from exploratory science and testing hypotheses in new regions. Discovery often happens when we put ourselves in conditions where we are likely to be surprised."

The Gakkel-Nansen Ridge forced major changes in the theory of how ridges form and evolve. Most other ridges have deep, cliff-walled rift valleys running along the top, offset frequently by perpendicular *transform faults*. But the Gakkel Ridge instead had a deep, central valley about 0.6 mile (1 km) deep and 31 miles (50 km) wide, flanked by staircase-like blocks of fractured rock along the walls.

Even more surprising, they found few signs of the transform faults that geologists thought came with all ridges and *spreading centers*. Those offsetting transform faults help pull the crust apart and generate fierce volcanic and rift activities on faster-spreading ridge systems, but the Gakkel Ridge formed a single, long, 1,100-mile (1,770-m) slash, without major offsets on connecting faults. The discovery posed a great mystery and challenged scientists to explain the unexpected findings.

ODD ROCKS LUBRICATE FISSURES

One clue lay in the discovery of large amounts of *serpentine*, a greenish, weak, easily crushed rock that is common in the mantle where it meets seawater. Perhaps the serpentine forms in the network of water-filled cracks (fissures) in the mantle's surface. This soft, malleable rock could then serve as something of a shock absorber along the fractures of the rift, preventing the shattering and displacement of the brittle crustal rock necessary to form the distinctive transform faults.

Those layers of weak serpentine might also account for the strange pattern of earthquakes along the ridge. Seismologists thought that the ridge was spreading apart slowly because the crust was thicker, and the molten energy of the magma pressing up from the mantle could not reach the surface. Therefore, they could expect fewer but more violent earthquakes as the thicker, colder slabs of crust stored up stress and then snapped.

Instead, they found exactly the opposite situation. Here the crust was wafer thin and laced with the soft, crushable layers of serpentine. So the thin, malleable crust absorbed stress without fracturing and caused unexpectedly weak earthquakes. Once again, the experts were completely surprised. However, the brush of the mantle of the Earth against the surface also produced far more volcanic activity than anyone expected, based strictly on the earthquakes and the rate of seafloor spreading.

So, once again, reality confounded the theories and predictions. Of course, such complete surprises make science exciting, even addictive. The back and forth between theory, experiment, prediction, and discovery makes science a mystery story that holds the answer to all the big questions.

7

Iceland

North Atlantic Ocean

At first, the tough, superstitious Icelanders figured it was just another *volcano*, rumbling, smoking, blasting, and finally spewing a furious, squeezed-out rush of *lava*. It was 1783, and they believed more in the spirits that lived in the *lava*, since the revolution of *plate tectonics* that would have turned superstition into science would not emerge for nearly 200 years. They did not know what lay in store for them, nor did they know that they lived on top of the only place where an undersea ridge breaks out into the sunlight.

The Icelanders had lived tenaciously for centuries on the world's most volcanically active piece of ground, a jagged jut of lava bigger than the state of Georgia, rising from the frigid waters halfway between ice-buried Greenland and frigid Scandinavia at the northern tip of Europe. Here steam rose routinely from hundreds of cracks in the Earth, and a dozen volcanoes muttered and sputtered and smoked at any given moment. During the last *Ice Age* 12,000 years ago, the scrap of land and the fuming volcanoes were all hidden under layers of ice miles thick. The ancient Greeks had heard stories about the remarkable island of steam and fire, which Phyheas of Massalia in 300 B.C.E. referred to as Ultima Thule, an island at the edge of the world six days north of Britain.

In the eighth century, a band of Irish monks set out in boats made of animal skins stretched over twigs to seek God and solitude on that semi-legendary land. They became the first settlers, living among the volcanoes and *geysers* that must have seemed to them the very face of God, speaking in thunder, smoke, and walls of molten rock. They were succeeded a century later by the Vikings, who established the first permanent colonies on the island.

From here, Eric the Red set out to discover Greenland, after he was exiled from Iceland for killing a man. He left with his clan and his followers who started a desperate, starvation-prone colony on the ice-plagued shores of Greenland. Icelanders also set foot on North

The research submersible *Alvin* explores the bottom of the Pacific Ocean during one of its record-breaking dives that revolutionized the understanding of the complex ecosystem created by volcanic activity along undersea ridges. *(National Oceanic and Atmospheric Administration)*

The *Trieste*, shown here at sea, has broken depth barriers and reached the deep-ocean floor by using lighter-than-water gasoline and ballast to rise and descend like a hot-air balloon. *(National Oceanic and Atmospheric Administration)*

The submersible *Alvin* descends toward the bottom of the Pacific, withstanding pressures that would quickly crush an ordinary submarine. *(National Oceanic and Atmospheric Administration)*

The same forces that created the Galápagos Rift built the volcanic Galápagos Islands, where the strange forms of birds, lizards, and other creatures that had evolved in isolation gave Charles Darwin crucial evidence to support his theory of evolution. *(National Oceanic and Atmospheric Administration)*

The robotic arm of a submersible research vessel collects a starfish from an undersea ridge in the Pacific Ocean. *(National Oceanic and Atmospheric Administration)*

Scientists were surprised to discover that tube worms also grow along fissures that leak hydrocarbons into the ocean in the Gulf of Mexico. *(National Oceanic and Atmospheric Administration)*

Tube worms make a living at the base of a black smoker in the Pacific Ocean. The worms survive right next to superheated water largely because the enormous pressure and frigid temperatures of the ocean bottom keep the heat contained. *(National Oceanic and Atmospheric Administration)*

Minerals venting from the seafloor. *(National Oceanic and Atmospheric Administration)*

The discovery of a "black smoker" such as this one on the Mid-Atlantic Ridge prompted many biologists to speculate that life on Earth may have originated at such a feature on the seafloor. *(National Oceanic and Atmospheric Administration)*

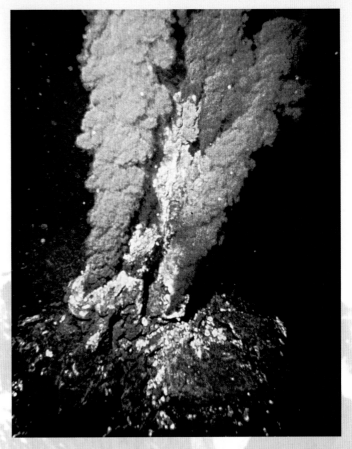

Mount Erebus in Antarctica fumes and smokes from a funnel-shaped crater, which conceals one of the few standing, lava-filled lakes in the world. *(National Oceanic and Atmospheric Administration)*

Cosmic rays produced by the Sun spiral down into the Earth's magnetic field creating the colorful displays of the aurora borealis near the North Pole. *(National Oceanic and Atmospheric Administration)*

Researchers must cope with deadly temperatures and dangerous ice flows to probe the secrets of the poles. *(National Oceanic and Atmospheric Administration)*

Volcanoes such as this one in Augustine, Alaska, mark the boundary between two crustal plates. *(National Oceanic and Atmospheric Administration)*

Wild horses roam the high, cold grasslands of Chile on a long, narrow valley running along the Andes Mountains, created by the subduction of a crustal plate into the Atacama Trench. *(National Oceanic and Atmospheric Administration)*

A sea turtle that flippers through the Red Sea, known to be an embryonic ocean. *(National Oceanic and Atmospheric Administration)*

Fish teem on coral reefs in the Red Sea, which lies in a rift between Africa and the Middle East. *(National Oceanic and Atmospheric Administration)*

THE FIRST ICELANDERS

In the ninth century, the Norwegian Vikings landed on the volcanic shores of Iceland, the first true settlers. They fled Norway to escape the bloody tyranny of Harald Fairhair. Igolful Arnarson arrived off the coast of Iceland in 870 C.E., tore apart his captain's chair, and tossed the heavy posts overboard. He vowed he would build a settlement wherever the gods caused those posts to wash ashore. He continued exploring the 200-mile- (322-km-) wide, 300-mile- (483-km-) long island for the next four years until he finally found the posts again, washed up on the volcanic shore in the southwest corner of Iceland in a bay so thickset with clouds he thought it was smoke rising from the innumerable *vents*. He called his settlement Reykjavik, or Smokey Bay.

The Vikings established a hardscrabble colony on the harsh shores of the volcanic island, drawing much of their sustenance from the sea and the struggling crops they planted in the volcanic soil. Hardy, independent, and living far from either civilization or tyrants, the Icelanders established the first democratic parliament in the world. Their democratic "Althing" first met in 930 C.E. beneath the glower of a sheer volcanic cliff, along a gigantic fissure that ran for miles through the heart of the island.

American soil long before Columbus. Leif Eriksson, nicknamed "Leif the Lucky," reached North America in 1000 C.E. before returning to Iceland. Columbus would not set foot on the continent for another 500 years.

The early settlers lived uneasily with the threat of cataclysm. In 935 C.E., the Eldgja Rift cracked open along a 20-mile (32.2-km) stretch. An awesome 20 cubic miles (83 km3) of lava flowed from the crack over the course of the next eight years. Fortunately, the population was sparse and not close to the rift, so it fed their sense of awe and mystery without causing serious damage.

The Icelanders held on tenaciously through the centuries, developing a rich mythology that incorporated the rumblings of the gods expressed through the constant turmoil of the ground beneath their feet. The Icelanders wove stories that populated the restless, angry, and capricious land with trolls, fairies, sprites, and spirits of the Earth and the ocean.

The Icelanders accepted hardship and loss as their inevitable lot. Outbreaks of disease among people who huddled closely together to survive the long winter remained common. So did crop failures and the resulting famines. In the 300-year span between 1400 and 1700, the population fell steadily to about 40,000, as rule passed to Norway and then to Denmark.

THE FURY OF THE EARTH SPIRITS

But even the Icelanders had not experienced the full fury of those spirits until the start of a three-year sequence of volcanic outpourings that started at Mount Lakagigar, the center of a rift in the Earth 15 miles

(24 km) long. Icelanders had grown used to short, sometimes violent explosions and generally slow-moving walls of lava that oozed out of cracks and rolled down to a furious, steaming meeting with the ocean. But this time, lava poured out all along the 15-mile gash in the Earth, accompanied by billows of a strange, poisonous, blue smoke. Even in far-off Europe, people turned to the west to note the odd-colored sky. The gas contained great billows of sulfur dioxide, the same element that nourished the hardy tube worms and bacteria of the vent communities on the ocean bottom. But in the air, the billows of sulfur dioxide proved poisonous as a rain of ash settled all over the island.

The awed and frightened Icelanders found themselves watching the greatest outpouring of lava in recorded human history. Other, much larger eruptions have taken place in the distant past, like a flood of lava 65 million years ago that covered 660,000 square miles (1.6 million km^2) in India. But this was the largest outpouring with known human witnesses.

The lava that covered Iceland in 1783, just as the Americans were struggling to start a new country, proved devastating for the Icelanders. Some three cubic miles of rock gushed out of the Earth, covering 224 square miles (580 km^2) and adding new land to the island. More important, the clouds of poison gas and the sulfur dioxide that settled into the soil killed most of the plants, which prompted the starvation of most of the livestock. Largely isolated from the rest of the world, perhaps one-third of the island's human population starved to death in the next two years.

The explosion offered a devastating lesson in the dangers of living on a mid-ocean ridge, since Iceland lies at the end of the Reykjanes Ridge, an offshoot of the Mid-Ocean Ridge. The volcanic island rose from the sea within the past 20 million years as a result of complex changes in the drifting *crustal plates* and the widening Mid-Atlantic Ridge. It forms a crucial link between the Mid-Atlantic Ridge and the much more complex fragmentation of ridges and crustal plates beneath the Arctic Ocean. It also forms the terrain for a fierce scientific battle that continues to shape our view of the Earth itself.

Geologists have been struggling for decades to crack the riddle of Iceland. The first detailed, continuous maps of the Mid-Atlantic Ridge that helped confirm the theory of *seafloor spreading* all pointed straight at Iceland. The island spews lava at a crooked elbow when the mostly north-south Mid-Atlantic Ridge turns to head toward the complicated ridge systems of the Arctic Ocean. In effect, Iceland is a northern section of the Mid-Atlantic Ridge that has risen above the ocean's surface, although the crest of the rest of the ridge remains miles beneath the surface in most places.

Geologists have struggled to explain why the Mid-Atlantic Ridge in Iceland rises so high above its average worldwide depth. The first and still leading theory suggested that Iceland represented the rare juxtaposition of the edge of a crustal plate and a *hot spot*, where molten rock torches the underside of the *crust* in a single spot.

A HOT SPOT OR SOMETHING ELSE?

Geologists speculated that Iceland was formed by the same sort of hot spot that built Hawaii. Such a heated plume of *magma* can melt the underlying crust and break through to the surface. But they believed that the hot spot that first built up the great bulge of Greenland was captured and locked into place when it hit this northern spur of the Mid-Atlantic Ridge, where it then created Iceland.

So geologists set out to map the seafloor and examine the rocks in Greenland and northern Canada to test the theory. They speculated that the hot spot started some 130 million years ago, just before the rift that split the Earth's *supercontinent* started to spread. The opening of that rift in the middle of the stomping ground of the dinosaurs turned a narrow *rift valley* into the Atlantic Ocean, splitting North and South America from Europe and Africa.

The rift gaped open and spread north until about 60 million years ago, when it split Scandinavia from Greenland and began opening up

THE DECCAN TRAPS

The terrifying outburst of lava from the semi-molten mantle that covered 660,000 square miles (1.6 million km^2) of west-central India with a thick layer of molten rock still puzzles geologists. Some argue that a giant asteroid punched through the crust and unleashed an outpouring of magma from the mantle. But most believe that a hot spot like a rising plume of rock from the mantle broke through the surface at a weak spot in the crust to unleash the outpouring of lava. Although the Deccan Traps now cover about 200,000 square miles (518,000 km^2), the original flow probably covered more like 660,000 square miles, with layers of lava up to 2,000 feet (610 m) deep.

Geologists speculate that perhaps the same sort of plume of rising, molten rock that has created a long chain of islands and undersea mountains that includes Hawaii came up under an existing plate boundary, fatally weakening the rift. Geologists have identified 40 or 50 such hot spots, including in Yellowstone National Park, Hawaii, and Iceland. For a long time, geologists blamed such hot spots on a single, mysterious weakness in the crust, but most now suspect the still poorly understood influence of *convection currents* at the boundary between the semi-molten mantle and the more deeply buried, fully molten, superheated *core*.

For instance, they note that the weak hot spot that created La Réunion Island in the Indian Ocean lay beneath India 65 million years ago. They argue that when that hot spot came up against the drifting edge of a crustal plate, it spurred the breakup of a continent and the greatest gush of lava in known history.

the Arctic Ocean, cracking the crust along the network of ridges east of Greenland. At that point, the hot spot that would eventually create Iceland was beneath Greenland, according to conventional theory. However, geologists fiercely debate the many lingering mysteries surrounding hot spots in general and the Iceland hot spot in particular.

At one time, most experts believed that hot spots started as a result of some weakness in the crust that allowed a great, elongated bubble of molten mantle rock to burn through the crust, perhaps initially blasting to the surface in a catastrophic eruption. Perhaps that is what created the *Deccan Traps* in India, they speculated. Perhaps such an initial "plugged" outburst of a hot spot plume built up the great, raised plateau of Iceland.

Many experts argue that the superheated, molten rock in the hot spot plume rises like a hot-air balloon through the cooler, compressed, much denser rock of the mantle. Picture a plume as resembling a blob of less dense oil rising in slow motion in a lava lamp. The Iceland hot spot could have created a 1,242-mile- (2,000-km-) wide "plume head" of molten rock that finally broke the surface along *fissures* and fractures.

Iceland has already helped confirm the theory that this rising body of less dense, molten rock drives hot spots. Recently, some geophysicists from the Georgia Institute of Technology put a supercomputer to work trying to study the mantle between 18 and 1,700 miles (29 and 2,720 km) beneath the surface. A team lead by geophysicist James Gaherty studied the movement of waves of energy generated by earthquakes. Such seismic waves move faster in hard, brittle rock than in semi-molten or molten rock. Seismic waves spread out in all directions from the epicenter of an earthquake when chunks of rock fracture and lurch past one another. Scientists studied the arrival time of the seismic waves from 17 earthquakes at different measuring stations. Since they knew exactly when the earthquake took place, they could measure the small differences in arrival times at the different stations. Using a sophisticated computer, they could then figure out the temperature and density of the rocks through which those seismic waves passed. They hoped that this complicated calculation would reveal the size and temperature of the plume of molten rock under Iceland.

Sure enough, they found a great plume of less dense rock rising from at least 60 miles (97 km) beneath Iceland, which meant it started out in the upper layers of the mantle. The plume of heated rock boosted the temperature in the surrounding rocks by about 175°F (79°C), apparently just enough to drive the volcanoes of Iceland. However, no one is sure what drives these hot spot plumes. Some experts speculate that perhaps they get started when one plate is forced beneath another. The leading edge of the crustal plate melts as it is forced miles beneath the surface.

THE HAWAII HOT SPOT

Hawaii remains the planet's best-studied hot spot, where molten rock from deep in the Earth escapes to the surface in a single spot as the crustal plates move overhead. Detailed mapping of the seafloor revealed that Hawaii is the youngest and highest of a 1,500-mile- (2,400-km-) long chain of 19 islands and numerous extinct, undersea volcanoes stretching halfway across the Pacific.

Like Iceland, the main islands that constitute the Hawaii Island chain grow continuously, still rising from the seafloor with the addition of new lava. Hawaii's Big Island is actually the tallest mountain on Earth, rising 56,000 feet (17,070 m) from the base of the seafloor to the top of Mauna Kea, which towers 13,796 feet (4,205 m) above sea level. By contrast, Mount Everest is about 28,000 feet (8,530 m) high.

Scientists believe that the superheated upwelling of a hot spot has melted a hole in the Earth's crust, causing lava to flow periodically to the surface. The stationary hot spot has spawned a long, connected chain of volcanoes as the restless crustal plates have moved over it. So the volcanoes caused by the hot spot mark the conveyor belt movement of the crust as basalt bubbles to the surface along the ridges and rifts that run along the East Pacific Rise.

Cemented ash coats the seafloor in Hawaii, one of a chain of islands formed as a stationary "hot spot" creates fissures in the bottom of crustal plates moving overhead. *(National Oceanic and Atmospheric Administration)*

This molten rock then oozes back up to the surface to create volcanoes and islands. Perhaps it also melts a fissure in the crust or finds the path of least resistance through which a hot spot can get started.

WHAT DRIVES HOT SPOTS?

Most experts now agree that these hot spots are driven by events 1,700 miles (2,736 km) below the surface, where the molten core of the Earth meets the semisolid rocks of the mantle. The molten rocks of the boiling core move in great convection currents, like a boiling pot of water. These convection currents press up against the mantle, causing related, slow-motion currents. Those indirect currents in the semi-molten mantle press up against the hard, cold, 15-mile- (25-km-) thick light rocks of the crust. In perhaps 40 known places scattered around the globe, that chain of currents creates a stable hot spot. Geologists have determined that these hot spots do wander slowly, but at only a fraction of the speed that crustal plates move on the conveyor belt moving outward from the ridges and down into the trenches.

Soon after the dinosaurs died and the Atlantic Ocean began to lurch open, Greenland began drifting northeast over what would eventually become the Iceland hot spot. The theory got a big boost from seafloor mapping, which showed a long, irregular ridge of undersea mountains running from the coast of Greenland straight to Iceland. Now as dead as the chain of volcanoes on the seafloor between Hawaii and Japan, the submerged Aegir Ridge looked like the smoking gun that proved Iceland was created by a stationary hot spot captured by the fissures underlying the Mid-Atlantic Ridge. Geologists estimate that the Aegir Ridge died 25 million years ago and Iceland first rose above the surface 14 to 20 million years ago.

However, several scientists have now challenged that theory, hypothesizing instead that Iceland is a result of the escape back to the surface of a huge piece of the Earth's crust that was buried, melted, and returned to the surface, rather than a hot spot fed by magma from the deeper mantle. Two different teams of scientists, one from the University of Durham and the other from the Geological Survey of Norway, have published their findings. First, they point out that the pattern of *magnetic stripes* on the seafloor that proved the Mid-Atlantic Ridge created the Atlantic Ocean offers a much more confusing picture on the seafloor between Greenland and Iceland. They insist that a hot spot could not have moved from Greenland to Iceland in the past 50 million years based on measurements of the seafloor magnetic stripes. Second, they point out that it would take two or three times as much lava to create Iceland and the surrounding undersea features as a single hot spot could reasonably produce. Third, the ages and chemistry of the rocks along the extinct Aegir Ridge are so different from Iceland that they could not have been created by the same hot spot.

Advocates of the hot spot plume have countered with different explanations. Perhaps there were multiple plumes, confusing the picture. Perhaps instead of a single plume there was a long rift that allowed hot mantle rock to bubble up along a line instead of in a single spot. Perhaps the superheated hot spot rocks rose until they hit fractures spreading outward from the ridge, so the molten rock was channeled along this system of cracks before rising to the surface.

But the dissident scientists have proposed another explanation entirely. They go back to the complicated bumping and grinding of crustal plates some 60 million years ago as *Pangaea* shattered. As Pangaea broke up and the bits and pieces went wandering off on their own, the rift that created the Mid-Atlantic Ridge reached the area of Greenland and Iceland. Here three different crustal plates smashed into each other. The scientists argue that some 60 million years ago the edge of one plate got forced underneath another, the same process that elsewhere has created the deepest places on Earth, undersea trenches. As the plates smashed together, the smallest of the three got trapped between the two larger, stronger plates.

So a huge chunk of the Earth's crust was pushed miles below the surface by the other, overriding plates. The deeper the buried crust went, the hotter it became. The rocks of the Earth have *radioactive* elements that continuously decay, giving off heat. This natural tendency of rock to generate heat keeps the core molten and raises the temperature of the rocks steadily with depth. So at some combination of depth, burial, and pressure, this buried piece of crust started to melt. As it turns out, the chemistry of light, silica-rich crustal rock is a little different from the denser rock of the mantle. This chemical difference could help account for the perplexing amount of lava that reaches the surface around Iceland. Moreover, a huge slab of melted, upwelling crust could seemingly generate lots more lava on the surface than a single, narrow plume melting its way up through the mantle and the overlying crust, they argued.

Geologists are still working out the contrasting theories, knowing that finding the truth in Iceland remains crucial in understanding the complex system of ridge, trenches, hot spots, and contending crustal plates that determines the shape of the face of the Earth.

ICELANDERS LIVE ON FIRE'S EDGE

Hardy Icelanders continue to live on the hard, heated crust of a geologist's dreamworld and a farmer's nightmare. Iceland is the only country on Earth that gets most of its energy from the Earth, since it has readily

> ## STRANGE LIFE ABOUNDS
>
> Remarkable creatures live in the ocean surrounding Iceland. One recent expedition used dredges and undersea robots to grab samples of the sea life living along the Mid-Atlantic Ridge system from Iceland to the Azores, a series of islands off the coast of Portugal that geologists determined was partially created by another hot spot. The expedition covered 40,000 miles (64,370 km), passing twice along the ridge whose upper reaches remain an average of 1.3 miles (2 km) beneath the surface.
>
> The biologists brought up 300 new species of fish and 50 new species of squid and octopus. That included one flaming red squid that lives its life in the absolute darkness a mile beneath the surface. They also discovered a strange, misshapen species of fish with needle-toothed jaws. The carnivorous fish sports fleshy, wriggling, wormlike appendages that jut out from the forehead and hang down toward enormous, gaping maws. Biologists also discovered a new species of anglerfish. All other known species of anglerfish are yellow and flat and hide on the bottom. This one was brown, bloated, and living happily in the mid-ocean. The discovery testified to the enormous adaptability of living things and the amazing way in which evolution can adapt the design of living things to new conditions.

harnessed the hot springs and geysers to run the economy. The people also enjoy the world's oldest-living tradition of democracy and one of the world's lowest crime rates. The steaming landscape remains steeped in myth, and it is perhaps the only place in the world where government engineers take into account the purported dwelling places of trolls and fairies when laying roadways. Perhaps a belief in spirits and myth flows naturally from living precariously on an ice-shrouded island with 24 active volcanoes.

Scenic Mount Hekla erupts about once every five years, coughing smoke and spitting ash to the delight of geo-tourists. The word *geyser* comes from the name of an explosive Icelandic hot-water spring that has since largely subsided. The island resembles a landscape designed by a crazy geologist. Rifts run for miles, punctuated by volcanoes. V-shaped cracks march along twisting paths. Desolate plateaus of jagged lava support strange varieties of moss, but remain too rough and jagged to walk across. One 150-foot (46-m) waterfall spilling over the edge of a sheer wall of lava generates a continuous glimmer of rainbows. The government at one time wanted to build a dam above the waterfall to generate electricity but abandoned the idea when a young woman threatened to throw herself over the falls if the government built the dam. The government relented and vowed to protect the falls.

The combination of fire and ice has created other dramatic effects. Some 8,000 years ago, the whole island was buried under a layer of ice several miles deep that connected it to an also-buried Greenland. The volcanoes continued to send molten rock steaming up against the bottoms of this vast ice sheet. The molten rock melted lakes inside the ice cap. Sometimes the eruptions would melt through the ice cap, unleashing the hidden lakes in titanic floods that have left their traces in the cataclysmic landscape of Iceland. As the Ice Age waned and many of the *glacier*s retreated, dramatic, steep valleys and sheer, flooded fjords were revealed.

Ice lakes still exist, since permanent glaciers cover about 10 percent of Iceland. The largest Icelandic glacier, Vatnajökull, covers 3,000 square miles (7,770 km^2) with a layer of ice up to half a mile deep. You could easily hide every single *glacier* in Europe in this single mass of Icelandic ice.

Meanwhile, the Icelanders continue to demonstrate their adaptability as they thrive on their little piece of the Mid-Atlantic Ridge with creative, fatalistic determination. Most heat their homes with steam and run their appliances with electricity generated by steam turbines built atop the geothermal vents.

The most recent major eruption tested that resolve and adaptability. Between 1963 and 1966, a whole new island rose out of the ocean as a result of the eruption of an underwater volcano. The newly forged Surtsey is one of 15 fresh islands fed from the same body of underground magma that drives the volcano Hekla. Some geologists think that these islands mark the opening of a new spreading center parallel to the existing plate boundary some 60 miles (97 km) west, the seed from which new oceans may grow.

VOLCANO THREATENS TOWN

But the biggest recent challenge to the optimism of Icelanders came in January 22, 1973, with the eruption of a brand new volcano, Eldfell. Black ash and lava poured out of the central crater, rolling down on the town and toward the narrow-necked harbor on which the fishing village depended. As the lava surged into the crater, the side of the mountain collapsed, sending a slow-moving wall of molten rock rolling toward the heart of town.

The fire department rushed to respond as the 5,000 residents of Heimaey fled. Soon the lava consumed the houses on the outskirts of town. But the firemen brought to bear every fire hose in town and turned a torrent of water on the molten rock. A desperate battle shrouded in billows of steam ensued as the firemen tried to cool the leading edge of the lava flow so it would form an instant rock wall and divert the rest of the lava into the ocean. Miraculously, they stopped the lava about one-fifth of the way through town, saving most of the buildings. But then the battle shifted fronts as the lava flowed furiously into the ocean. The townspeople realized that the lava could easily block the narrow entrance to the harbor, killing the fishing village just as effectively as if it had burned every house in it. So the firemen shifted their lines and again doused the leading edge of the lava. Remarkably, they once again stopped the lava. Interestingly, the flow of molten rock flowed out into the ocean just far enough to significantly improve the harbor.

Once again, the resourceful people of Iceland proved that you can live on a mid-ocean ridge.

8

The Java Trench
Indian Ocean

Tilly Smith played happily in the placid, turquoise-blue waters of a beach in Phuket, Thailand, on December 26, 2004, determined to forget about school and textbooks and tests. She was 10, a world away from her home in England, and freed for the moment from the demands of school. But as she stood looking out at the Indian Ocean, something very strange happened. The tide started to suddenly draw water away from the beach. Something from her geography class tugged at her memory, something very scary about when the water draws away from the beach. Suddenly, she remembered. A *tsunami!* Her teacher had said an approaching tsunami draws water away from a beach for perhaps 10 minutes before it arrives. Suddenly terrified, she looked up and down at the beach at the 100 or so happy tourists, many of them turning now in surprise to watch the water draw away from the beach. "Mummy," she cried. "Mummy, we must get off the beach now!" Her mother looked at her, puzzled. "There's going to be a tsunami," said the little girl, stumbling at the word. Several adults standing nearby stared at her, then back out to sea, then once again at the little girl. She turned and began running away from the ocean. Her mother hesitated a moment, then began shouting, "tsunami, tsunami, get away from the beach." Others took up the cry and soon hundreds of people were running as fast as they could away from the beach.

JAVA TRENCH (SUNDA DOUBLE TRENCH) VITAL STATISTICS

- 190 miles (306 km) off the coast of Sumatra and Java in Indonesia
- 24,440 feet (7,450 m), maximum depth
- 1,600 miles (2,575 km) in length

GIRL'S GEOGRAPHY LESSON SAVES LIVES

Tilly's geography lesson saved hundreds of lives. "Last term [geography teacher Andrew] Kearney taught us about earthquakes and how they can cause tsunamis," Tilly told a newspaper reporter. "I was on the beach and the water started to go funny. There were bubbles and the tide went out all of a sudden. I recognized what was happening and had a feeling there was going to be a tsunami. I told mummy." She did not know it until much later, but her moment of insight and quick action provided one of the few bits of good news in what would soon prove one of the great natural catastrophes in human history. The people on the beach with Tilly all escaped the onrushing tsunami, but more than 150,000 other people on coastlines all around the Indian Ocean proved tragically less lucky.

Some two hours before Tilly noticed the water drawing away from the beach, a 600-mile (960-km) section along the deep-buried Java Trench had ripped loose, unleashing a 9.0 earthquake centered beneath the seafloor off the coast of the Indonesian island of Sumatra. The

The massive Alaskan earthquake of 1963 left fishing boats scattered like toys in the streets of the devastated city. *(National Oceanic and Atmospheric Administration)*

inexorable pressure on the descending Pacific plate had finally overcome centuries of friction and resistance deep beneath the surface and so plunged deeper down into the Earth. As a result, a 600-mile-long chunk of the overlying plate suddenly lurched upward by about 15 feet (4.6 m).

That sudden movement of the Earth miles beneath the ocean's surface lifted the water above the seafloor. Because water cannot be compressed, the energy of the earthquake was transmitted up through the ocean toward the surface. The energies involved were tremendous. The force of the earthquake created an open ocean swell of water one to four feet (1.2 m) in height. That does not sound like much, but a cubic yard (27 cubic feet or .76 m^3) of water weighs a ton, and this vast bulge in the ocean quickly covered thousands of square miles of water.

The wave spread outward from the epicenter of the quake beneath the complex of trenches that includes the Java Trench. By the time it reached the surface, it was spreading outward at nearly 500 miles (805 km) an hour, the speed of a jetliner.

Ships out at sea barely noticed the swell of the gathering tsunami as it swept past, bearing down on the islands that fill the Indian Ocean. A network of wave detectors would have recorded the onrushing wave, but the impoverished, seismically quiet Indian Ocean nations had no such network. Pacific nations, including the United States, have a network of floating sensors anchored to the bottom of the ocean. Alerted by an earthquake, these sensors can detect the passing pressure wave of a developing tsunami and bounce an alarm off a satellite within minutes. Scientists would then send out an alert, giving emergency officials in coastal communities precious hours to evacuate the beaches.

Unfortunately, the people on beaches from Africa to Asia had no warning at all. The wave rose up out of the ocean as it neared land. It hit with the greatest fury along the eastern cost of Sumatra, rising to 40 or 80 feet (12 or 24 m) and the height of a six-story building in places. The

THE WORLD'S WORST TSUNAMIS

- 2004: Indian Ocean near Indonesia, 150,000 dead
- 1755: East Atlantic, 60,000 dead in Lisbon alone
- 1883: Krakatoa near Indonesia, 36,000 dead
- 1896: Honshu, Japan, 27,000 dead
- 1976: Philippines, 5,000–8,000 dead
- 1960: Chile, 2,300 dead
- 1906: Ecuador and Colombia, 500–1,500 dead
- 1868: Chile, 25,000 dead

wave killed more than 100,000 people in Indonesia, completely obliterating many villages and smashing one or two miles (1.6 or 3.2 km) inland in places.

The waves spread outward from that epicenter on the northern tip of Sumatra. It raced north, killing some 5,000 people in Thailand, where Tilly's parents were vacationing.

Julie and Casey Sobolewski, Americans who had chartered a sailboat for a tropical vacation in Thailand, saw the wave coming from half a mile offshore outside of Rai Leh Beach. They were heading toward a sandy spit of beach where 150 tourists lolled in the bright sunlight. The Soboleskis watched in horror as a 30-foot- (9-m-) tall wave of water swept over the beach and swallowed everyone on the beach. The wave rampaged toward them, splintering five small fishing boats and hurtling everyone on them into the now furious water. The Soboleskis turned the boat toward the wave, realizing they could never outrun it.

Somehow, the 35-foot- (11-m-) long sailboat managed to climb the face of a wave as tall as it was long and break through to the other side without overturning. They spent the next several hours plucking swimmers out of the water.

Faye Linda Wachs and Eugen Kim, from California, were scuba diving in 120 feet (37 m) of water when the wave washed over them. They felt a sudden, strong surge, and the water grew abruptly murky, but the wave overhead had not yet touched bottom and so had not curled and broken. Bewildered, they swam back to shore. To their horror, bodies soon started floating past them in the now cloudy water. As they emerged from the water, they found a scene of chaos and devastation, their hotel and all their possessions consumed by the ocean. One palm tree had a speedboat impaled upside-down on its crown. "There were lots of broken legs and deep gash wounds like you'd expect to see in the Civil War," Wachs told

DISASTER WARNING SYSTEM

Although the ridges and trenches running beneath the Indian Ocean generate fewer earthquakes than their counterparts in the Pacific, geologists knew the potential for tragedy. For instance, a 9.2 earthquake in 1833 created a tsunami similar to the 2004 disaster. That rupture in the Earth took place on the same system of trenches a few hundred miles south of the epicenter of the 2004 quake.

Fortunately, that shift in the Earth generated a wave that headed southwest into open ocean rather than blasting the coastlines to the north and west. Still, just weeks before the tragic 2004 quake, an Australian geologist wrote an article about the 1833 quake and warned that the nations around the Indian Ocean should consider installing a tsunami warning system to mirror the system protecting the nations of the Pacific.

the *New York Times*. "We began seeing we had just freakishly survived a natural catastrophe. It was a weird, horrible nightmarish 'Survivor' situation," said Kim.

WAVE EXTRACTS TERRIBLE TOLL

The wave raced across the Indian Ocean, bearing down on millions of unsuspecting people. It killed more than 30,000 in Sri Lanka and 10,000 in India, but it was not finished. It continued west. The wave inundated every square inch of the Maldives Islands, where the highest point is only five feet (1.5 m) above sea level. Miraculously, the wave killed perhaps 100 people there, although it left the entire population homeless. The tsunami continued on to the coast of Africa, smashing into Somalia, Kenya, Tanzania, and Madagascar. That terrible wave ultimately killed more than 150,000 people and left millions homeless, making it one of the costliest and most tragic natural disasters in history. But it did not come as a surprise to geologists who had been studying the Java Trench,

A tsunami comes ashore with tragic consequences. *(National Oceanic and Atmospheric Administration)*

part of a network of deep folds in the ocean floor caused by the slide of one crustal plate under another.

The network of trenches running along the edge of those vast *crustal plates* has repeatedly generated earthquakes and tsunamis. Moreover, the same geological forces that create these disasters also built the islands scattered across the Indian Ocean, thereby extracting a tragic price for their previous gifts.

That same system of trenches that generated the worst tsunami in recent history also generated the most devastating volcanic explosion in history, the cataclysmic eruption of Krakatoa, which generated a firestorm and tsunami that killed an estimated 36,000 people.

All these disasters are rooted in the mysterious depths of the Java Trench, along with a network of similar deep gashes in the Earth that includes the Mariana, Philippines, Ryukyu, and Palau Trenches.

TRENCHES MIRROR RIDGES

The deep-sea trenches are the mirrors to the undersea ridges. The *crust* manufactured at the ridges is recycled back into the Earth beneath the trenches. Think of the crust of the Earth as a conveyor belt, rising up from the semi-molten *mantle* along the fracture of the ridges, hardening as it emerges onto the cold, dark seafloor. More *magma* pushing up from behind moves the newly formed crust away from the ridge, which is what created the pattern of *magnetic stripes* that allowed geologists to unearth the theory of *plate tectonics*. The light, silica-rich rock of the continents "floats" on top of this seafloor conveyor belt. But what happens to all that crust manufactured at the ridges, then shouldered roughly out of the way?

Most of it ends up in the bottom of a deep-sea trench. Inch by inch, over the course of millions of years, this jigsaw puzzle collision of crustal plates—with their embedded, floating continents—will bump up against another crustal plate. Sometimes the two plates plow into one another head-on, their edges crumpling like minivans in a slow-motion head-on collision. The edges crumple, raising stupendous mountain ranges like the Himalayas. Sometimes they jostle, shift, and slide past each other sideways, creating offsetting faults like the *San Andreas Fault*, which runs the length of California. Often continental plates paste on mountain ranges in such a sideswipe, like the Sierra Nevada that divides California from Nevada. On the other hand, one plate sometimes gets forced underneath the other, creating an undersea trench along the boundary between the two plates, one riding on top of the other. The trenches form the downward-dipping, destructive final process that started with the system of expanding undersea ridges.

BURIED PLATE MELTS

As a chunk of cool, light, rigid continental crust is forced deep down into the Earth by the pressure of the overlying plate, it heats up—becoming hotter the deeper it goes. The heat is generated partly by the friction between the two plates and partly by the heat from the decay of naturally occurring *radioactive* elements present in the Earth that builds up with depth and explains why the Earth's mantle is semi-molten and the *core* entirely liquid. Down where the hard, cooler *crust* gives way to the semi-molten upper reaches of the mantle, it is hot enough to melt rock and power *volcano*es. The buildup of friction between the plates generates most of the world's earthquakes, which effectively outline the edges of the crustal plates all around the world (see color insert on page C-7 [top]).

The trenches all form in the seafloor when one plate plunges underneath another. Once that leading edge of the buried plate melts, the now superheated, less dense melted rock will rise to the surface along any weakness in the overlying rock. As a result, the world's deep-sea trenches all have an echo of volcanoes, mountain ranges, and island arcs within a

A volcanic fumarole fumes in the Aleutian Islands. *(National Oceanic and Atmospheric Administration)*

few hundred miles of the trench, where the melted leading edge of the buried trench escapes back to the surface. For instance, the Peru-Chile Trench hugs the coast of Peru where the Nazca plate dives down under the South American plate. The melted edge of the buried Nazca plate eventually escapes to the surface in a chain of volcanoes that built the towering Andes Mountains. Near Alaska, the Aleutian Trench runs along the northern edge of the Pacific. Here the Pacific plate drops down under the North American plate. The leading edge of the buried plate melts and rises, creating the long, thin chain of the Aleutian Islands that extends from the edge of North America toward Russia. The Java Trench features some of the most complex and dynamic geological impacts of this *subduction* of a crustal plate into an undersea trench. The Java Trench has also spawned an exotic, complicated, endangered jostling of islands.

The Pacific plate dives under several other plates, creating a wound of trenches plunging five to seven miles beneath the surface. Those buried plate edges in turn create numerous island chains, including the Aleutians, the Kamchatka Peninsula, Japan, the Philippines, Borneo, Sumatra, Java, New Guinea, and New Zealand. Four different plates collide along that long zigzagging arc of trenches, including the Pacific, Australian, Indian, and North American. In addition, several smaller plates get caught up in the chaos, including the Philippine and Caroline plates, remnants of larger plates now trapped in this collision of giants.

That complex jostling of plates has created a jigsaw mystery of crushed plates, deep voids, furious volcanoes, and devastating earthquakes that has shaped not only the evolution of life but also the rise and fall of past human civilizations.

Geologists have spent decades trying to unravel it all. They have made detailed maps of the seafloor, folded, wrinkled, crunched, and ripped by the slow impact of the plates. Some plates get swallowed, some hit a glancing blow and spin, and some slide to the side.

Scientists have undertaken thousands of studies to understand the dynamics of this collision of crustal plates near undersea trenches. They have undertaken detailed analysis of the rocks and *lava*s. They have mapped thousands of earthquakes to trace plates' edges. They have also studied changes in the speed of energy waves generated by earthquakes as they pass through the Earth deep beneath the trenches, where the buried edges of the plates heat up, melt, and begin to form new rock. Seismic waves produced by earthquakes travel at different speeds through the cold crustal rock, the semi-molten mantle, and the molten core. So scientists use powerful computers and hundreds of seismographs scattered all over the planet to measure precisely the arrival time of energy waves generated by distant earthquakes. Nearby seismographs will measure waves that have passed through the crust. However, the

The Katmai volcano forms a perfect crater. *(National Oceanic and Atmospheric Administration).*

energy waves that reach the far side of the world have to pass through the mantle. So geologists can use the changing travel times to measure exactly where the crust gives way to the mantle. Moreover, they can chart where the leading edge of one crustal plate descends down underneath another behind a trench.

Seismologists can use the earthquake energy waves to get a picture of what is happening miles beneath the surface. Along earthquake faults such as the San Andreas in California, two great slabs of the Earth lurch slowly past each other along a strike-slip fault. But behind undersea trenches where the edge of a plate plunges down for melting and recycling, earthquakes generate a very different kind of energy.

Earthquakes caused by the ongoing recycling of the Pacific plate along that rim of fire continue to cause devastation, as they have in this region for all of human history. A significant earthquake occurs on an average of once per day along the long, violent edge of the Pacific plate. In the Indonesian area, ongoing earthquakes and the tsunamis they cause killed at least 20,000 people between 1990 and the 2004 eruption.

For instance, in 1994, a sudden lurch, miles beneath the ocean surface near Java, produced a 7.8 Richter scale earthquake. The quake generated a tsunami that savaged the coast of Java with waves of up to 36 feet (11 m), killing 200 people. Of course, that quake is dwarfed by the staggering tsunami of 2004, which had waves that could have been 90 feet (27 m) tall in some places.

HISTORY'S WORST VOLCANO

Another of history's terrible tsunamis was also a by-product of the titanic forces that formed the Java Trench. Eyewitnesses have left vivid descriptions of the eruption of Krakatoa (Krakatau) in 1883. For instance, Captain Lindemann of the passenger ship *Loudon* had loaded up with sightseers on a volcano tour on that fateful day. One of 130 active volcanoes in

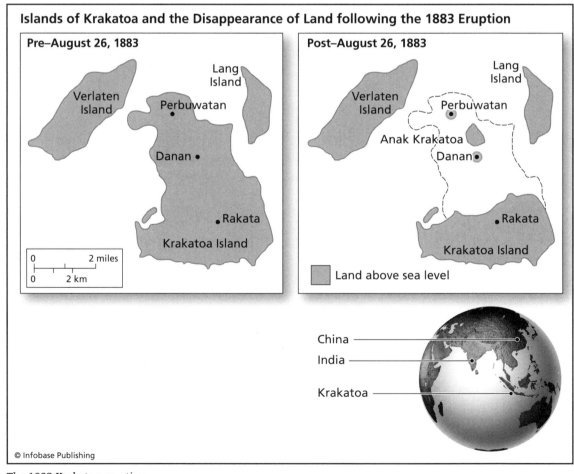

The 1883 Krakatoa eruption

the innumerable islands that make up Indonesia, Krakatoa had rumbled to life on May 20, 1883, spewing a cloud of smoke, ash, and *pumice* some eight miles (13 km) into the sky. The small, uninhabited island was dominated by three merged volcanoes, rising from the ocean to a height of perhaps 1,200 feet (366 m). Lying in a strategically vital, narrow straight between the tropical volcanic islands of Java and Sumatra, Krakatoa had collapsed explosively back in 416 C.E., leaving a five-mile- (8-km-) wide crater. But in the ensuing 1,500 years, regular volcanic explosions had rebuilt the island.

No one understood the source of volcanoes at that time. No one knew that the five-mile- (8-km-) deep Java Trench lay just west of the volcano and that the same forces that had created the trench had also created all those volcanic islands, laying the groundwork for the tragedy that was about to take place along this boundary between the Indian Ocean and the Pacific Ocean.

The reawakening of Krakatoa initially spurred more interest than alarm. Sightseers chartered boats to get close to the island and marvel at the fury of black smoke rising far into the sky, filled with flecks of volcanic rock. The *pumice* filled with air bubbles as it cooled, drifting down in a rain of ash and floating on the surface of the ocean. Already, huge clumps of this ash were floating in the narrow shipping lanes. Geologists who sampled the pumice noted that it was composed of basalt, which means that magma from deep in the Earth was being fed into the magma chambers below the surface. Three volcanoes had grown separately then merged to create a single island about seven miles (11.3 km) long and less than four miles (6.4 km) wide. All three volcanic peaks and perhaps eight other major *vents* all began furiously coughing ash and pumice. No one knew enough about volcanoes at that time to understand how ominous was the appearance of basalt fed to the surface from deep down near the Earth's mantle.

The eruptions were accompanied by earthquakes. But the quakes were centered deep under the seafloor, miles to the south, beyond the still unplumbed depths of the Java Trench, part of a chain of vast canyons running in an intermittent arch from New Guinea nearly to Asia.

But even if they had known enough to understand the warning signs and the titanic forces driving the volcano, no one would have imagined the next, terrible stage. At about 1 P.M. on August 26, an ear-shattering explosion sounded in the volcano, and within moments, the plume of ash rose from about six miles (9.7 km) high to nearly 20 miles (32 km) high. In the next few hours, the outpouring accelerated, until it reached a height of close to 30 miles (48 km), injecting light, air-filled flecks of ash and pumice high into the atmosphere where the ever-present, planetary wind currents of the jet stream begin to spread it across the planet. Even that was only the warning shot.

At about 5:30 A.M., the entire island experienced in a shattering succession of four explosions that dwarfed anything any living human being had witnessed, then or since. In a matter of minutes, two-thirds of the island was blown into the sky, eventually converting the 1,200-foot- (366-m-) high, three-peaked mountain into a four-mile- (6.4-km-) wide, 200-foot- (61-m-) deep crater. Some 15 square miles (39 km^2) of the island were either vaporized and hurled into the sky or dropped into the suddenly empty crater that had once fed magma into the volcano. People 3,000 or 4,000 miles (4,800 or 6,400 km) away reported hearing the explosion, which sounded all the way from Australia to Sri Lanka.

But the worst was yet to come. The explosion and the rush of rock into the sea and the collapse of the remains of the island into that suddenly created four-mile-wide crater instantly created massive tsunami waves. Waves up to 120 feet (37 m) tall moved outward from the shattered island.

Captain Lindemann had just recovered his composure as he moved out of Lompong Bay in Sumatra when he saw the monster wave advancing toward his small ship, powered by a steam-fired paddle wheel on the side. He was nearly 50 miles (80 km) from the volcano, so it never occurred to him that the massive explosion he had heard posed a direct threat to his ship.

"Suddenly we saw a gigantic wave of prodigious height advancing toward the seashore with considerable speed. Immediately, the crew managed to set sail in face of the imminent danger; the ship had just enough time to meet with the wave from the front. The ship met the wave head on and the *Loudon* was lifted up with a dizzying rapidity and made a formidable leap. The ship rode at high angle over the crest of the wave and down the other side. The wave continued on its journey toward the land and the benumbed crew watched as the sea in a single sweeping motion consumed the town. There, where an instant before had lain the town of Telok Betong, nothing remained but open sea." Another steamship still in the harbor was picked up and deposited a mile inland, 30 feet (9 m) above sea level, killing the entire crew of 28.

The waves killed many people and ripped loose vegetation on many small, low-lying islands. For instance, the wave swept over Sebesi Island to the northeast, stripping away all plants and washing away the 3,000 doomed inhabitants. Even 60 miles (97 km) from the volcano, residents of places like Thousand Islands had to climb trees to escape the wave that covered even the highest places on the island.

On the big islands of Sumatra and Java, the 100-foot- (30-m-) tall waves swept far inland before breaking against the towering volcanic peaks that had in their turn built the islands. One man working in a rice field five miles (8 km) from the shore in Java left this haunting account:

All of a sudden, there came a great noise. We saw a great black thing, a long way off, coming towards us. It was very high and very strong and we soon saw that it was water. Trees and houses were washed away. The people began to run for their lives. Not far off was some steep, sloping ground. We all ran towards it and tried to climb up out of the way of the water. The wave was too quick for most of them and many were drowned almost at my side. There was a general rush to climb up in one particular place. This caused a great block and many of them got wedged together and could not move. Then they struggled and fought, screaming and crying out all the time. Those below tried to make those above them move on again by biting their heels. A great struggle took place for a few moments, but one after another, they were washed down and carried away by the rushing waters. You can see the marks on the hillside where the fight for life took place. Some dragged others down with them. They could not let go their hold, nor could those above them release themselves from this death grip.

No one knows how many people died in those first few hours of the greatest volcanic explosion in historical record, but estimates range from 20,000 to 36,000. The tsunami accounted for 90 percent of the deaths. But Krakatoa was still not finished.

KRAKATOA: WHEN LAVA FLIES

The massive explosion of Krakatoa created a bizarre, superheated "pyroclastic flow" of ash and steam that incinerated seaside villages miles from the volcano. The collapse of the volcano instantly vaporized a huge quantity of rock and ocean water. More gas than solid, this superheated mixture of molten rock and ash reached the ocean and rushed, headlong, across the surface. These pyroclastic flows traveled an astonishing 30 miles (48 km) across open water to blast the shoreline of nearby Sumatra.
One woman in the village of Ketimbang in southern Sumatra recalled the horrifying experience:

Suddenly it became pitch dark. The last thing I saw was the ash being pushed up through the cracks in the floorboards, like a fountain. I turned to my husband and heard him say in despair "where is the knife? I will cut all our wrists and then we shall be released from our suffering sooner." The knife could not be found. I felt a heavy pressure, throwing me to the ground. Then it seemed as if all the air was being sucked away and I could not breath. I felt people rolling over me. No sound came from my husband or children. I tottered, doubled up, to the door. I tripped and fell. I realized the ash was hot and I tried to protect my face with my hands. The hot bite of the pumice pricked like needles. Without thinking, I walked hopefully forward. Had I been in my right mind, I would have understood what a dangerous thing it was to plunge into the hellish darkness. I entangled myself more and more. I noticed for the first time that skin was hanging off me everywhere, thick and moist from the ash stuck to it. Thinking it must be dirty, I wanted to pull bits of skin off, but that was still more painful. I did not know I had been burnt.

ASH AFFECTS CLIMATE

Krakatoa blasted at least 20 times as much ash and pumice into the atmosphere as did the eruption of Mount St. Helens in the United States nearly a century later. Ash rained from the sky 2,000 miles (3,200 km) from the volcano and choked the narrow straights all around so that open ocean looked like solid ground. The finest particles of ash and stone reached the *stratosphere*, along with a massive injection of sulfur dioxide gas. As a result, acid rained from the sky all over the planet in ensuing weeks.

Moreover, the fine flecks of ash that reached the upper atmosphere spread clear around the planet on the winds of the jet stream within two weeks and remained suspended there for years. The volcanic dust actually blocked part of the Sun's energy, causing a drop in the temperature of several degrees across the entire planet. For years after the explosion, the ash high in the atmosphere resulted in unusually colorful sunrises and sunsets across 70 percent of the planet. Eerie halos also appeared around the Sun and Moon.

In 1927, to the terror and astonishment of a Japanese fisherman tending his nets nearby, the tip of a new volcano poked above the waves from the center of the four-mile- (6.4-km-) wide, 200-foot- (61-m-) deep crater, or caldera. A tiny island spewing steam from its round, central crater has now risen from the ashes of that disaster. The residents of the nearby islands, with no living memory of that terrible explosion, dubbed it Anak Krakatoa, or Child of Krakatoa.

Among geologists there are still differences of opinion about why Krakatoa created such a shattering explosion. Most geologists believe the volcano was fed by a massive chamber filled with magma from deep in the Earth. At some point, either the magma retreated or a breech in the volcano allowed this vast underground chamber to abruptly empty out.

THE WORLD'S WORST EARTHQUAKES		
COUNTRY	YEAR	DEATHS
China	1976	242,000
China	1920	180,000
Japan	1923	143,000
Soviet Union	1948	110,000
Italy	1908	75,000
China	1932	70,000
Peru	1970	66,794
Pakistan	1935	60,000
Iran	1990	40,000

As a result, the whole island fell into the pit, along with a rush of seawater, which hit the magma beneath the chamber and instantly vaporized. The collapse of the caldera could explain the violence of the explosion. On the other hand, some geologists suggest that perhaps magma rushed upward from deeper in the mantle, simply blasting off the island like a cork in a champagne bottle. As evidence, they point to the basalt in the ash that blasted out of Krakatoa.

The tragic events of 2004 underscored the titanic energies still at work beneath those trenches. And though no one has ever ventured to the bottom of the Java Trench with anything but robot cameras and *sonar* signals, its path and dynamics still hold millions hostage. And as Tilly discovered, it also underscores the value of geography classes and knowing when to start running.

9

Peru-Chile Trench

Pacific Ocean

Geologists Simon Lamb and Paul Davis wanted to answer a vexing question: Why were the towering Andes Mountains along the coast of Chile and Peru rising so fast? The answer surprised them and offered a startling illustration of how an undersea trench can have unexpected impacts on the number of fish in the ocean, birds in the rain forest, and even global weather patterns.

The question posed by Lamb and Davis eventually led into the depths of the Peru-Chile (or Atacama) Trench that hugs the coast of South America less than 100 miles (161 km) offshore. The trench runs for 3,666 miles (5,900 km), making it the longest submarine trench on the planet. It plunges to a depth of 26,460 feet (8,065 m) in Richards Deep and averages about 40 miles (64 km) in width. That gives it an area of 228,000 square miles (590,500 km²). But those dimensions are not the most impressive things about the Atacama Trench.

Geologists have long marveled at the sheer wall of rock that forms the Andes Mountains, the landward echo of the Atacama Trench (see the color insert on page C-1 [bottom]). The rise from the 26,460-foot depths of the trench to the 19,700-foot (6,005-m) peaks of the Andes remains the sharpest vertical rise on the planet, an altitude gain of 40,000 feet (12,190 m) in a distance of less than 50 miles (80 km).

Those mountains continue to rise at a dramatic pace. For instance, Charles Darwin, whose revolutionary theory of evolution was developed on the nearby *Galápagos Islands*, also studied birds along the coast of Chile. He noticed odd piles of broken-open shells that certain seabirds left on flat places next to the shore. These piles of bird-gathered shells were dozens of feet above the high-tide line. Clearly, the coast had risen significantly before the shells could even decay or get buried.

Experts on *plate tectonics* can now explain what Darwin observed about the rising elevation of the islands. Out in the Pacific off the western coast of South America, the long, low East Pacific Rise splits the seafloor.

> ## THE TSUNAMI OF 1868
>
> The slow-motion crash of *crustal plates* that created the Atacama Trench and built the Andes Mountains has also repeatedly spawned death and tragedy for the people who live in the shadow of those events.
>
> One of the most devastating events involved a wave triggered by an 8.5 magnitude earthquake centered where the South American and Nazca plates lurched over one another.
>
> Strain had accumulated on these two opposing crustal plates for centuries. Eventually, the force pressing the two plates overcame the friction and resistance of the two slabs of *crust* far beneath the surface. A huge section of the Earth lurched, transmitting the force to the ocean overhead. The movement of the plates created a 90-foot- (27-m-) tall wave that spread outward from near the Atacama Trench at the southern tip of Chile. The massive wave sped down on the port city of Arica, where two American warships and one Peruvian ship rode at anchor. Minutes after the quake shattered many buildings in town, the approaching wave eerily drew much of the water out of the harbor as it advanced. A moment later, the wave arrived in a fury. The wave smashed one warship against a harbor island, killing all but two of the crewmembers. The wave also snapped the moorings of two other ships. Floating free, the two ships rode the wave far inland, then went rushing back toward the harbor as the wave withdrew. One ship was crushed, losing almost all of its 83 person crew. But the other ship wound up sitting upright, almost unharmed on the beach 1,290 feet (393 m) from the water.
>
> The earthquake and tsunami killed as many as 70,000 people along the South American coast. It also crossed the Pacific Ocean to send 15-foot (4.6-m) waves crashing ashore in Hawaii and New Zealand and five-foot (1.5-m) waves in Japan.

Along this rise, scientists first discovered the undersea *vents* near the Galápagos Islands. The upwelling of *magma* into the East Pacific Rise drives the Pacific plate apart from the much smaller Nazca plate to the east, forcing it up against the South American plate. The two plates crash into one another at a faster rate than any other plate boundary on the planet, driven by the East Pacific Rise's spreading rate of seven inches (18 cm) per year. The long undulation of the Atacama Trench is formed where the small, dense, 3.8-mile- (6-km-) thick Nazca plate plunges under the lighter, 25-mile- (40-km-) thick South American plate. That slow-motion collision has shaped the spectacular geology of Chile and Peru, plus the five-mile- (8-km-) deep trench just offshore. As the Nazca plate descended into the Earth, the melted bits of crust fed *volcanoes* and triggered frequent, devastating earthquakes. Measurements of earthquake *epicenters* suggest that the buried plate is melting and breaking up deep beneath the surface.

The two plates first collided with one another about 200 million years ago as the *supercontinent Pangaea* was splintering and fragmenting. The complex geology of the Andes reflects the prolonged event. The oldest rocks crumpled and tilted in its peaks date back 250 to 450 million years ago. But most of the rocks are the much younger outpouring of volcanoes. The Andes still boast the world's highest volcanoes.

The Andes now have two, jagged, parallel lines of peaks separated by a high, flat valley filled with rock and sediment eroded off the surrounding mountains. This central valley provides almost the only farmable land in the whole, rugged country. At one time, it sustained the remarkable and resourceful Inca, who supported a larger population than lives there today.

Studies of the rocks and gases still spewing from those volcanoes have revealed connections to the deepest levels of the Earth. Geologists suspect that the volcanic plumbing of the Andes goes clear down to the Earth's *mantle*. Lamb has traveled all over the Andes, collecting gas samples from springs and vents. In fact, one vent nearly scalded him to death moments after he took a sample, he reported in his book *Devil in the Mountains*. He discovered traces of helium in the vent, probably directly from the Earth's mantle. Helium is a simple, almost indestructible molecule. Once, nearly the whole universe was made of hydrogen and helium. The fusion reactions of stars have converted these light elements into all the other heavier elements that make up the universe.

COLLISION OF PLATES CREATES ANDES

The long, violent crunch involving the two plates under the Atacama Trench created the Andes. Rocks scraped off the downward-plunging Nazca plate piled up against the toe of the South American plate, creating a gigantic bulge as the foundation for the towering Andes.

This description of the collision of the two plates made sense and fit in perfectly with existing theories. However, the experts could not quite explain the speed with which the Andes had risen. They were puzzled by the discovery that the mountains were surging upward faster than any other mountains in the world. Moreover, they did not understand the strange unevenness of that rise.

As they pondered the mystery, Lamb and Davis tried to calculate the shear stresses between the two crustal plates beneath the trench. They reasoned that the friction between those two plates should determine how rapidly the mountains would rise. They knew those calculations would also help explain why the area sometimes produces devastating earthquakes and waves. Usually, strain builds up on those buried pieces of crust until the strain overcomes the friction between the rocks of the two plates. The result is an earthquake that raises or lowers the mountains.

Here is where the geologists got an exciting surprise. They noticed that the largest earthquakes occurred directly opposite stretches of the Atacama Trench that had the most mud in the bottom. Intrigued, they

The Andes Mountains continue their rapid rise, pushed upward as the result of the subduction of one crustal plate under another along the fissure of the Chile Trench. *(National Oceanic and Atmospheric Administration)*

shifted the focus of their study. They compared the rate of mountain building to the sediment in the opposite section of trench and found a surprising fit. What could it mean?

Different sections of the Atacama Trench fill with mud washed off the nearby continent at dramatically different rates. At the southern end, several rivers rush down from the Andes and empty their load of mud into the ocean. At that southern end, the trench has nearly filled up with miles of mud. However, the towering wall of the Andes rises almost directly from the coast along much of the Atacama Trench. One 800-mile (1,287-km) stretch of coast has hardly any rivers at all. Along these stretches of the trench, little sediment reaches the ocean. In this section, the trench plunges to five miles (8 km) deep, with only a thin layer of sediment washed down from the mountains. Here the mountains are also the highest and the earthquakes most infrequent.

The evidence led Lamb to a startling conclusion. The mud washed down into the bottom of the trench actually helped lubricate the boundary between the two plates, even miles beneath the surface. And that raised another startling realization.

CLIMATE CAN AFFECT MOUNTAIN BUILDING

The climate can actually affect the rise of a chain of mountains. For instance, today the western slope of the Andes gets almost no rain. The moisture-laden breezes off the ocean hit the wall of the mountains and produce monotonous, soupy fogs half the year. However, they rarely rise high enough to bring rain to the mountains. As a result, the central valley between the Andes is a high, cold desert. Certain areas have gone 17 years without a trace of rain.

Ironically, the wall of the Andes also largely accounts for the heavy rainfall on the opposite side of the range. That means the Andes play a key role in creating the remarkable diversity of the Amazonian rain forest in Brazil. Moisture-laden air from the Atlantic Ocean gets blocked by the Andes and so provides the massive rainfall that sustains the rain forests. Those rain forests are the lungs of the planet, absorbing carbon dioxide and releasing oxygen. The rain forests also harbor more than half of the planet's living *terrestrial* species, all protected by the wall of mountains built by the presence of the Atacama Trench.

The sometimes treacherous waters of the Pacific exact a toll on the fishermen who make their living in one of the world's most productive fisheries, driven by a combination of ocean currents and upwelling of nutrient-rich water from the Chile Trench. *(National Oceanic and Atmospheric Administration)*

> ## STUCK IN A TRENCH
>
> When a continent embedded on one crustal plate hits the trench that sometimes forms between two plates, the continental mass of rock can actually get stuck in a trench and "flip" it. Imagine what happens when two continental plates smack into each other and one gets pushed down or subducted under the other. Usually, the older, heavier plate will get pushed down under the lighter, younger plate, forming a trench perhaps 70 miles (113 km) wide and 1,000 miles (1,609 km) long.
>
> Now suppose the plate that dives down underneath is carrying a continent made of light, silica-rich rock that "floats" on the dense basalt of the oceanic crust. The light continent will start to go down into the trench, but then get stuck in the throat of the trench because it is so light. Initially, the leading edge crumples and folds and scrapes off some of the oceanic crust from the overlying plate. Eventually, the light rock of the continent will block the descent of the plate on which it is embedded.
>
> However, the pressure continues to build, since both plates are still being jammed against each other. Eventually, the plates will "flip," and the plate with the continent will start to ride up over the other plate. Sometimes the trench will then disappear, to be replaced by a great mountain range.

But Lamb and Davis soon realized that past changes in climate might have affected the amount of sediment washed down into the trench and therefore the *lubrication* between the crustal plates. In short, warm, wet, rainy periods would slow mountain building by providing more lubrication, and dry, low-erosion periods would build higher mountains faster.

But that is not the only intriguing link between the Atacama Trench, climate, and biology. Consider the impacts of the ocean currents, the atmosphere, the diets of fish, and the vital weather and sea surface interaction known as "El Niño." Normally, the mysterious, frigid, lightless depths of the Atacama Trench play a vital role in making the waters off the coast of Chile a paradise for fish. The upper, sunlit stretches of the ocean nourish most of the life in the ocean, thanks to *photosynthesizing*, microscopic plankton, the ocean's equivalent of grass. The whole food chain depends on plankton. Plankton floats in the upper layers, turning sunlight into energy, just like plants do on land. When plankton die, they sink to the bottom, unless they are devoured first.

TRENCHES ACCUMULATE NUTRIENTS

In shallow waters, a host of creatures live at the bottom, sustained by this rain of nutrients from above. However, few life-forms can survive in the lightless deserts in the bottoms of the trenches. Hot springs and vents support a complex group of creatures, but few species live in the trench bottoms away from the vents. The brief visit of the *Trieste* to the bottom of the Mariana Trench revealed some signs of life, but nothing to compare to either the vent communities on the ridges or the shallower depths near the continents. As a result, the plankton that drifts down into the trenches from the sunlit layers does not get eaten by bottom dwellers. Instead, it accumulates, which makes the cold, deep waters of the trenches rich in nutrients.

Peru volcano

> ## SEARCH FOR THE ORIGINS OF THE ANDES
>
> Geologist Simon Lamb has spent much of his life in the Andes, looking for the "devil in the mountain." At least, that is the name of his absorbing book published by Princeton University Press about his adventures in the high, strange places of the Andes Mountains. He has spent years climbing over a barren land trying to understand the forces that thrust the Andes up some 40,000 feet (12,190 m) from the depths of the adjacent Peru-Chile undersea trench. He named his book for the Bolivian god Tio, who guards the mineral treasures of the mountains. Ironically, geologists have found that minerals such as gold, silver, and copper are forged by the superheated interaction of the ocean and magma along the *spreading centers* of undersea ridges. The riches of the Andes were thus forged in the ocean, then thrust up by the collision of two crustal plates to the hidden places of the Andes.
>
> Lamb has explored the remote peaks and valleys of the Andes, often coming across the ruins of the Inca civilization at altitudes of 12,000 to 19,000 feet (3,658 to 5,852 m), where the air is so thin most people gasp for oxygen after only modest exercise. He has survived landsides, poisonous volcanic vents, rockfalls, disease, hardship, and frostbite, all to unravel the mysteries of the mountain.

Normally, prevailing winds blow down off the Andes and out to sea, driving strong surface currents. These surface currents create an upwelling from the depths of the Atacama Trench. This cold water carries with it the rich, nutrient soup of the plankton that has rained into the depths. So the combination of the nutrient-rich, cold waters of the trench and the wind patterns makes the waters off the coast of Chile exceptionally rich in nutrients. Fish flock from all over the ocean to take advantage of these floating nutrients, especially sardines. And that nourishes a whole food chain of predators, such as tuna, and one of the world's best fisheries.

Usually, the waters off the coast of Chile above the Atacama Trench are cold and rich. But every two to seven years, warm water suddenly sweeps in close to the coast, driven by a circulation change that sends climatic ripples across the whole planet. The warm water cuts off the upwelling from the trench. Abruptly, the herring, anchovies, squid, and other cold-loving creatures dwindle or vanish. That spells disaster for the fishermen, sea birds, sea lions, and other creatures that depend on the now vanished fish. Meanwhile, the warm surface waters also generate a deluge of rainstorms that batter the Andes and the coast of Chile. And that increases the flow of mud into the trench, which can perhaps affect the speed with which the mountains rise.

This pattern of periodic warm surface currents off the coast of South America has been dubbed El Niño. The cycle has persisted for thousands of years, as demonstrated by the way in which the Inca built their cities and irrigation canals. Clearly, they adapted to these strange cycles of drought and flood. But the importance of El Niño spreads far from the edges of the Atacama Trench. No one is quite certain what triggers El Niño. It apparently starts with the failure of the normally dependable trade winds that usually blow along the equator. The decline of those winds tends to signal the start of El Niño. Gradually, the normal wind patterns reverse, so weaker trade winds blow in the opposite direction. Soon warm surface water from the western Pacific spreads east, gaining strength as it cuts off the cold upwelling from the Atacama Trench.

This spread of warm water across the middle of the Pacific generates wet clouds and storms in the eastern Pacific and southern North America. It also decreases the number of Atlantic hurricanes.

This simple change in sea-surface temperature can have startling impacts. For instance, one terrible drought in India between 1789 and 1793 caused by an El Niño pattern led to the deaths of an estimated 600,000 people through crop failure and starvation. El Niño increases the odds of wildfires throughout the world and droughts in Indonesia,

The normally abundant fisheries off the coast of Chile sometimes fail when the warm El Niño currents bring warm, nutrient-impoverished water up against the coast of South America. *(National Oceanic and Atmospheric Administration)*

> ## THE NITRATE MYSTERY
>
> Geologists have long puzzled at the extensive deposits of nitrogen in the otherwise barren and waterless Atacama Desert, in the high, dry, cold valley running between the two long ridges of the Andes Mountains.
>
> How did nitrates produced by living things get scattered in the soil of 4,000 square miles (10,360 km^2) of frigid, nearly waterless desert? Surprisingly, they came from deep in the Atacama Trench, a 3,600-mile- (5,790-km-) long, five-mile- (8-km-) deep trench 100 miles (161 km) off the coast of Chile. Plankton living in the ocean use the Sun's energy to take nitrogen out of the atmosphere and convert it to forms other creatures can use. When the microscopic plankton dies, it sinks down into the depths of the Atacama Trench. But eventually, winds blowing off the shore of Chile drive surface currents that cause upwelling and bring this deep, cold, nitrogen-rich water to the surface, sustaining one of the world's richest fisheries. That plankton-produced nitrogen gets caught up in the fogs that form regularly along the coast. In the course of millions of years, winds have carried traces of that nitrogen-rich fog up and over the desert high in the Andes.

eastern Australia, New Guinea and West Africa, and northern South America. The 1982–83 El Niño caused floods and typhoons that killed 2,000 people in the United States, Peru, Ecuador, Bolivia, Cuba, Hawaii, and Tahiti and drought elsewhere. Estimates put damages at a total of $13 billion. All of which demonstrates the unexpected and vital role that undersea trenches and ridges play in seemingly unrelated events.

As it turns out, trenches can affect the weather and, surprisingly enough, the weather can affect trenches. This discovery demonstrates how science works and the unexpected impact of a great idea. Alfred Wegener started out a century ago just trying to put the Humpty Dumpty of the continents back together again. But his creativity and stubborn pursuit of the truth despite the ridicule of other scientists helped spawn this revolution in our understanding of the world. Suddenly, it all fits together. So what lies in store? How will the pieces get shuffled next?

10

Red Sea

North Africa and the Middle East

The nascent undersea ridge opening in the Red Sea in the Middle East provides a glimpse of the titanic processes that long ago broke up the vast *Pangaea supercontinent* to create the continental outlines of the modern world. The narrow Red Sea and the connected Gulf of Aqaba form a dynamic arm of Africa's Great Rift Valley, which forms the beginnings of a new ocean and a new world. The world's next great undersea ridge is beginning to emerge along that long arch, which is just the latest step in dismantling the single supercontinent that nurtured the dinosaurs before it started breaking up some 300 million years ago.

A fissure began to gape in northern Africa some 35 million years ago, opening up along the straight line of the 1,600-mile- (2,570-km-) long Red Sea, with a maximum depth of more than 7,500 feet (2,290 m). The widening gap split Arabia off the African continent. The Red Sea is really an offset continuation of the Indian Ocean Ridge. It may also mark the terrain that spawned modern humans, as evidenced by fossils discovered in Africa's Great Rift Valley. This rift has produced the geography that was the landscape on which Western civilization developed.

Sea horses live in the teeming coral reefs of the Red Sea, which now forms a nascent ocean that is pushing Africa and the Middle East apart. *(National Oceanic and Atmospheric Administration)*

Here the sagging, rumbling *crust* is being stretched and split, just as the Atlantic Ocean opened up as a narrow rift valley some 200 million years ago in the midst of Pangaea, the equator-straddling supercontinent.

Many geophysicists maintain that the Red Sea is really a young ocean that will open to become the next Atlantic Ocean, separating Africa from *Eurasia*. (A sea turtle flippering through this "young ocean" can be seen in the color insert on page C-8 [top].) The crack that created the Red Sea also runs through the Gulf of Aqaba. The Red Sea is the deepest part of a system of rifts and troughs that will eventually rip loose a huge chunk of eastern Africa. Geologists think that the ridge emerging under the Red Sea will split into a system of ridges dividing at least three different *crustal plates*. That will eventually create an ocean ridge running down Africa's Great Rift Valley. However, other geologists suspect shifts by other plates will instead seal up this third arm of the rift before it opens.

The nearly 4,000-mile- (6,440-km-) long rift system remains complex and poorly understood. Some sections are splitting apart. Other sections are instead *transform faults*, like the *San Andreas Fault* in California that offsets two colliding plates as they slide past each other laterally. The entire system starts near Syria and runs down the famed Jordan River, along the Sea of Galilee, through the Dead Sea, and finally to the Red Sea before continuing down Africa. The portions of the system along this stretch are offsetting transform faults. The actual crack does not start rifting until well into Africa, where the complex crustal edge splits into two rift systems.

The rifted slump in the crust is marked in Africa by a series of mountains, such as *glacier*-graced, volcanic Mount Kilimanjaro. The rift has also created deep troughs filled with some of the world's largest freshwater lakes. Several plateaus divide the two rift systems, and here titanic geological forces have created 4,500-foot- (1,370-m-) deep Lake Tanganyika and Lake Victoria, the second-largest freshwater lake in the world.

Today these deep rift valleys are deserts, dry and remote. But it was here that archaeologists found the 3-million-year-old skeleton of "Lucy," one of the earliest human ancestors. Experts on human evolution now believe that several contending species of small, upright-walking, big-brained humanoids lived in the forests and savannas of East Africa. They spread outward across the other continents from there, probably in several waves. Modern *homo sapiens*, the descendants of Lucy and her kind, spread outward from Africa perhaps between 100,000 and 200,000 years ago, after their long incubation in that rift valley straining now to become an ocean.

RED SEA NOURISHED CIVILIZATION

Oddly enough, these same titanic forces are responsible for the geography that nurtured the rise of Western civilizations and three of the world's major religions. The warm, enclosed Red Sea, with water temperatures a balmy 72°F (22°C) a mile (1.6 km) down, has nourished dreamers, mystics, ancient civilizations, and trading empires. Deep hot springs saturated with minerals were visited by researchers from Woods Hole Oceanographic Institution in the 1960s. They discovered rich, 300-foot- (91-m-) thick seafloor deposits of iron, *manganese*, zinc, and copper more than a mile beneath the surface. Those *vents* are rich in *sulfide* and poor in oxygen, just like the vents steaming from the seafloor along other mid-ocean ridge systems. Geologists believe that the minerals were forged by superheated water seeping through the cracks deep beneath the sediment-filled Red Sea.

To the north of the Red Sea, the Suez Canal connects directly to the Mediterranean Sea. While the Red Sea is opening, the Mediterranean is slowly closing as the jostling of the plates shifts Africa away from the

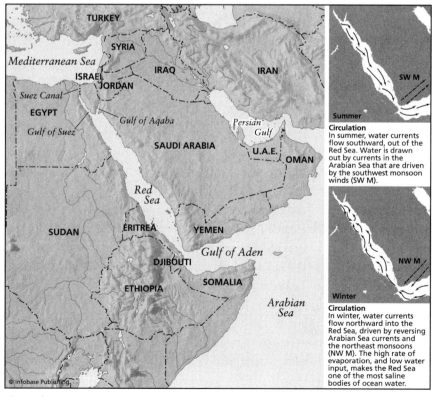

The Red Sea

Two fish hunt for something to eat in the rich ecosystem of the Red Sea's coral reefs. *(National Oceanic and Atmospheric Administration)*

Middle East (fish teeming on coral reefs in a rift between Africa and the Middle East can be seen in the color insert on page C-8 [bottom]). Africa will eventually rotate and plow into Europe, perhaps raising up a crumpled mountain chain to rival the Himalayas.

The split drove Arabia into Asia, pushing the small, dwindling Arabian plate down underneath Asia. That formed a trench, but the collision is nearly complete so that all that remains is a shallow sea. Sediment soon filled in the basin. As the sea level rose and fell with shifts in the climate, the ocean repeatedly invaded the shallow basin. Every time the water evaporated, it left another layer of salt. Eventually, these salt layers sealed up the sediment trapped in the bottom of the basin.

Those buried marshes and swamps capped by thick layers of salt created the vast oil fields of the Middle East. That oil fuels the world's economy, not to mention the tragic succession of wars that have plagued the region in the 20th century.

Earth will continue to shift and change, driven by the creation of new crust along the system of ridges and the consumption of old crust along the echoing arch of trenches.

University of Texas geologist Christopher Scotese has tried to predict the intricate dance of the continents in the next 250 million years. Intriguingly, his computer model predicts that by then they will reassemble a single, new supercontinent he has dubbed "Pangaea Ultima." The game of continental bumper cars could go something like this, with each continent floating along at the speed your fingernails grow: Africa will pivot and move north, smashing into Europe and raising a massive mountain chain of crumpled crust where the Mediterranean is now. That could leave a new Mount Everest, topped with bits and pieces of seafloor. Australia will catch its breath and make a dash for Asia, plowing over that great scattering of Pacific Islands, smashing into Japan and plowing into the northern coast of Asia to create yet another cataclysm

A red sea fan filters its food from the water on a coral reef in the Red Sea. *(National Oceanic and Atmospheric Administration)*

of mountain building. *Antarctica* will trail along after, but shift and come careening into the toe of India. Meanwhile, North America will initially drift farther from Europe as the Mid-Atlantic Ridge continues to widen the Atlantic Ocean by pushing out more crust.

Scotese admits that his geological crystal ball gets pretty cloudy after about 50 million years into the future. "It's like you're traveling on the highway, you can predict where you're going to be in an hour, but if there's an accident or you have to exit, you're going to change direction," Scotese said in an interview with Firstscience.com. "And we have to try to understand what causes those changes. That's where we have to make some guesses about the far future, 150 to 250 million years from now."

One big question centers on mysteries deep down in the undersea trenches. The most basic theory of *plate tectonics* suggests that the whole system is driven by *convection currents* in the Earth's *mantle*. Those currents rise from the heat source of the *core*, like the convection currents in a pot of boiling water. When the rising, hotter current hits the underside of the hard, cooler crust, it melts a long fracture and creates a mid-ocean ridge. The current in the mantle then flows along the underside of the crust, dragging along the overlying crustal plate. But eventually the semi-molten rock down in the mantle cools and becomes denser, then sinks back down toward the core. Observations show that downward-dropping current creates a *subduction* zone that forms a trench and forces the crust back down with it.

The Earth's surface is divided into perhaps 20 different crustal plates whose distributions do not line up neatly with a simple pattern of convection currents in the mantle. The scattering of trenches is even harder to fit into the simplest model. Trenches mark only some crustal edges opposite ridges. Sometimes plates just crumple up against one another. Sometimes they are smashed out of existence between several neighboring plates. Sometimes the edges shift sideways into long transform faults.

CREATING A NEW TRENCH?

Some geologists now think that trenches may form as a result of a weakness in the crust. Maybe they form near *hot spot*s like the one that built Hawaii. Perhaps the weight of the crust and the embedded continents causes the crust to sag and make a trough. Perhaps tilted crust can create a trough in the middle of a crustal plate, rupturing the crust and then plunging down into the Earth with the momentum of its own weight. This "slab-pull" theory has gained support and represents a challenge to the standard "river raft" theory, which suggests the continents are riding along on a conveyor belt of ocean crust.

Here Scotese parts company with many other geologists. He sees evidence of the development of a new trench in the middle of a crustal plate. He predicts this trench will open up in the western Atlantic and cause plates to shift direction. That could put an end to the growth of the Atlantic Ocean.

Of course, most geologists would disagree with that prediction. They believe that trenches can only form along the boundary between two crustal plates, where the crust is already stretched, fractured, and weakened. But his projection suggests that North America will pivot on the rifting of a new trench system and move north. In that case, North America will come plowing down on the Arctic Ocean and smash into the northern reaches of *Eurasia*, once again, raising a Himalayan-size ruckus.

Scotese predicts that after 250 million years the continents will wind up more or less right back where they started, a single landmass centered on the equator. The remains of the Indian Ocean would create a gigantic, landlocked sea in the middle of Pangaea Ultima. Alaska would stick out like a giant thumb at the northern tip. And then the process would start all over again. In fact, Scotese's reconstruction suggests that the continents do a dance that lasts roughly 500 million years, repeatedly gathering together, splitting up, and then reassembling as the Earth's plates grow.

However, his projections also suggest that the land area grows a little bit with each cycle. That happens because the light rock of the continent generally floats on the ocean crust, without getting pulled down into the trenches. The tendency of the light continental rock to float on the oceanic crust without getting pulled into the trenches explains why some rocks on the continents are 4 billion years old while the oldest rocks on the seafloor are not much more than 200 million years old. *Volcano*es constantly add rock to the continents, and collisions between plates sometimes also heave great chunks of the seafloor up onto the raft of the continents. So that means that the light continents tend to grow, without losing much of their substance to the recycling of crust in the trenches. Therefore, the total area of the continents may grow from cycle to cycle.

In the meantime, the dynamics of undersea ridges and trenches will continue to shape the history of life on the planet. Some 250 million years ago, the first dinosaurs were preparing for a long, brilliant run as evolution's top dogs. In fact, mammals did not get well launched until some 65 million years ago, once a mysterious cataclysm cleared the dinosaurs from life's stage, perhaps a terrible fresh start for life ordered up by the same titanic forces that run trenches and ridges. Our first, humanlike ancestors did not emerge as a distinct species until perhaps 4 million years ago, and human beings did not even leave Africa until maybe 100,000 years ago. So the 250 million years before the continents have their reunion is a long time in the future.

Moreover, predicting the future of the Earth demands educated guesses, imaginative scientists, and hard work. Scotese and the other geologists are still trying to answer some of the very questions that drove Wegener to propose his groundbreaking idea. They have made astonishing discoveries in the past century—*magnetic stripes*, incredible depths, new forms of life, fuming volcanoes, and primordial gases. But each question so ingeniously, painfully, brilliantly answered only calls up a host of new questions, like earthquake swarms spreading outward from every shift in the intellectual crust of science.

In the end, that is how science works and why human beings have at least started to understand how the world is put together. All you need to be a scientist is the next great question—and the strength, will, and discipline to seek the answer.

Glossary

anomalies departures from the normal or common order, form, or rule

Antarctica a continent within the Antarctic Circle asymmetrically centered on the South Pole. Some 95 percent of Antarctica is covered by a mile (1.6 km) thick icecap. The region was first explored in the early 1800s, and although there are no permanent settlements, many countries have made territorial claims

axis a straight line about which a body or geometric object rotates or may be conceived to rotate

bathysphere a reinforced, spherical deep-diving chamber suspended by a cable and used to study the oceans. The bathysphere, limited to depths of about 3,000 feet (900 m), was supplanted by the safer and more navigable bathyscaph

Bering Sea a northward extension of the Pacific Ocean between Siberia and Alaska that lies north of the Aleutian Islands and connects with the Arctic Ocean through the Bering Strait

bioluminescence emission of visible light by living organisms such as the firefly and various fish, fungi, and bacteria

black smoker an undersea vent that releases superheated water so rich in sulfur compounds that it billows what appears to be black smoke

buoyant a tendency to float due to a lower density than the surrounding gas or fluid

Challenger Deep the deepest portion of the Marianna Trench, the deepest of all undersea trenches

chemosynthesis the synthesis of carbohydrate from carbon dioxide and water using energy obtained from the chemical oxidation of simple inorganic compounds. This form of synthesis is limited to certain bacteria and fungi

compression the act or process of compressing or squeezing a substance into a smaller volume through the application of pressure

convection cells   circulation patterns caused by heating a fluid from below, including magma movements in the Earth's molten core that ultimately drive seafloor spreading and continental drift

convection currents   heat transfer in a gas or liquid by the circulation of currents from one region to another. The convection currents in the molten core of the Earth are thought to cause the movement of the crustal plates at the surface

core   the central portion of Earth below the mantle, beginning at a depth of about 1,800 miles (2,900 km) and probably consisting of iron and nickel, including a solid inner core and a liquid outer core

crust   the exterior portion of the Earth that lies above the Mohorovicic discontinuity, a density layer at which energy waves change velocity. The outermost solid layer of a planet or moon

crustal plates   the layers of cool, light, brittle rock that "float" on top of the semi-molten rock of the Earth's mantle

crystallize   to form crystals or assume a crystalline structure.

Deccan Traps   a portion of the plateau of south-central India between the Eastern Ghats and the Western Ghats that is covered with a massive volcanic outpouring closely associated with a planetary mass extinction

electromagnets   a magnet made from a coil of insulated wire wrapped around a soft iron core that is magnetized only when current flows through the wire

epicenters   the point of the Earth's surface directly above the focus of an earthquake

Eurasia   the landmass comprising the continents of Europe and Asia

fissures   cracks, including slits in the seafloor caused by volcanic and seismic activity

Galápagos Islands   Pacific Ocean volcanic islands west of the mainland of Ecuador. The islands are famous for their rare species of fauna, including giant tortoises. Charles Darwin visited the islands in 1835 and collected a wealth of scientific data that contributed to his theory of natural selection

geyser   a natural hot spring that intermittently ejects a column of water and steam into the air

glacier   a huge mass of ice slowly flowing over land, formed from compacted snow in an area where snow accumulation exceeds melting and sublimation

gravitometer   an instrument used to measure specific gravity or an instrument used to measure variations in a gravitational field

helium a colorless, odorless, inert gaseous element occurring in natural gas and with radioactive ores. It is used as a component of artificial atmospheres and laser media, as a refrigerant, as a lifting gas for balloons, and as a superfluid in cryogenic research

hemoglobin the iron-containing pigment in red blood cells of vertebrates that carries oxygen and consists of 6 percent heme and 94 percent globin

Homo sapiens the present-day species of human beings, the only existing species of the primate family Hominidae

hot spot an area of intense heat, radiation, or activity. Also a concentration of heat below the crust that creates a chain of volcanoes at a fixed spot as the thin, cool crust moves past overhead. Famous hot spots include Hawaii and Iceland

ice age a cold period marked by periods of extensive glaciation alternating with episodes of relative warmth

land bridges connections between continents, generally referring to land that dries out during ice ages when ocean levels drop

lava molten rock that reaches the Earth's surface through a volcano or fissure, or rocks formed by the cooling and solidifying of molten rock

lemur any of several small, tree-dwelling, mostly nocturnal primates chiefly of the family Lemuridae of Madagascar and adjacent islands, having large eyes, a long slim muzzle, and a long tail

limpets marine gastropod mollusks, including the families Acmaeidae and Patellidae, having a conical shell and adhering to rocks of tidal areas

lubrication a substance, such as grease or oil, that reduces friction when applied as a surface coating to moving parts

magma the molten rock material under the Earth's crust, from which igneous rock is formed by cooling

magnetic field magnets, electric currents, and the convection currents in the molten core of the Earth create a field characterized by the existence of a detectable magnetic force at every point in the region and by the existence of magnetic poles

magnetic pole flip at irregular intervals the Earth's magnetic poles flip, so that compasses will point south instead of north. No one knows what causes the pole flipping, but scientists suspect it has to do with currents in the molten core of the Earth

magnetic stripes a pattern of magnetic polarity in cooled lava on the seafloor close to the spreading center of an undersea ridge. Magnetic elements in the lava align themselves with the north magnetic pole, and the strips of cooled lava on the seafloor form magnetic strips as the magnetic pole of the Earth periodically reverses itself

magnetometer   an instrument for measuring the intensity and direction of a magnetic field

manganese   a gray-white or silvery brittle metallic element found worldwide, especially in the ores pyrolusite and rhodochrosite and in nodules on the ocean floor. Mixed into steel, it increases strength, hardness, wear resistance, and other properties

mantle   the layer of the Earth between the crust and the core

metabolism   the physical and chemical processes in a living cell or organism necessary for the maintenance of life as some substances are broken down to yield energy, while other vital substances are synthesized

ocean basins   the low-lying areas between continents, usually bounded by systems of trenches and ridges

olivine   a mineral silicate of iron and magnesium, principally $(Mg, Fe)_2SiO_4$, in igneous and metamorphic rocks

Pangaea   the supercontinent into which most of the Earth's landmasses were gathered during the Triassic, before the movement of the crustal plates caused it to break up and the continents to move to their current positions

periodite   a group of igneous rocks composed mainly of olivine and various pyroxenes with a texture like granite

photosynthesis   the process in green plants and certain other organisms by which carbohydrates are made from carbon dioxide and water using light as an energy source. Most forms of photosynthesis release oxygen as a by-product

pillow lavas   pillow-shaped balls of lava that ooze out onto the ocean floor where the frigid, high-pressure water causes its instant hardening into a distinctive shape

plate tectonics   the theory that focuses on the division of the Earth's surface into a mosaic of jostling crustal plates as the explanation for such diverse features as undersea ridges, movements of continents, undersea trenches, volcanoes, and earthquakes

pumice   a light, porous, glassy lava, used in solid form as an abrasive and in powdered form as a polish and an abrasive

radioactive   using or derived from the energy of atomic nuclei, caused by the spontaneous decay of particles

rift valley   a deep fracture or break, about 15 to 30 miles (25 to 50 km) wide, extending along the crest of a mid-ocean ridge

San Andreas Fault   a zone of crustal fractures extending along the coastline of California from the northwest part of the state to the Gulf of California. Movement of the tectonic plates along the fault has caused numerous tremors, including the devastating San Francisco earthquake of 1906

seafloor spreading   the split between two crustal plates fed by magma from the mantle is forced toward the surface. Seafloor spread takes place along undersea ridges

seamounts   isolated undersea mountains built by an extinct volcano that started near a spreading center, ridge, or trench and then moved out away from the volcanically active area. Also known as a guyot

serpentine   any of a group of greenish, brownish, or spotted minerals, $Mg_3Si_2O_5(OH)4$, used as a source of magnesium, asbestos, and decorative stone

sonar   a system using transmitted and reflected underwater sound waves to detect and locate submerged objects or measure the distance to the floor of a body of water

spreading centers   areas on the seafloor where two crustal plates are forced apart by magma welling up from below

Sputnik   a series of Soviet satellites sent into Earth orbit, especially the first, launched October 4, 1957

stratosphere   the region of the atmosphere above the troposphere and below the mesosphere

strike-slip faults   an earthquake fault in which one side moves up and often on top of the other, instead of sliding past one another as in a transform fault

subduction   a geologic process in which one edge of a crustal plate is forced below the edge of another

submersible   a vessel that operated or remains underwater

sulfide   a compound of bivalent sulfur with an electropositive element or group, especially a binary compound of sulfur with a metal

supercontinent   the gathering of most of the continents into a single landmass as a result of seafloor spreading. Such supercontinents gather and break up in the course of hundreds of millions of years

terrestrial   relating to Earth or its inhabitants or relating to or composed of land

thermocline   a layer in a body of water that sharply separates regions differing in temperature, creating an abrupt temperature gradient across the layer

transform faults   an offset section of a fault in which the two sides of the fault move past one another, like the San Andreas Fault

transponders   instruments that emit a signal or pulse, for use in things like marking a position on the seafloor

tsunami   a very large ocean wave caused by an underwater earthquake or volcanic eruption

vents an opening permitting the escape of fumes, a liquid, a gas, or steam, as in the volcanic vents discharging superheated water on the seafloor in zones of seafloor spreading

volcano an opening in a planet's crust through which molten lava, ash, and gases are ejected or mountains formed by the materials ejected from a volcano

Books

Broad, William J. *The Universe Below.* New York: Simon & Schuster, 1997. Excellent general book on the oceans by the science writer of the *New York Times.*

Colby, Carroll. *Underwater Exploration under the Surface of the Sea.* New York: Putnam Publishing Group, 1966. Good basic history of early expeditions.

Couper, Alastair. *Atlas and Encyclopedia of the Sea.* London: Times Books, 1992. Great maps with good information.

Cox, Allan. *Plate Tectonics: How It Works.* Oxford, England: Blackwell Scientific, 1986. Basic explanation of the physics of the process.

Day, Trevor. *Oceans.* New York: Facts On File, 2006. Excellent general overview of the topic for middle and high school students.

Earle, Sylvia A. *Atlas of the Oceans.* Washington, D.C.: National Geographic, 2005. Beautifully illustrated maps of the oceans loaded with excellent information.

———. *Sea Change: A Message of the Oceans.* New York: G. P. Putnam's Sons, 1995. Strongly argued look at ocean's environmental problems.

Ellis, Richard. *The Search for the Giant Squid.* New York: The Lyons Press, 1998. A fascinating account of the attempts to locate this semi-mythical beast.

Firor, John. *The Changing Atmosphere: A Global Challenge.* New Haven, Conn.: Yale University Press, 1990. A detailed look at the intricate connections between the oceans and the atmosphere.

Fogg, G. E., and David Smith. *The Explorations of Antarctica.* London: Cassel, 1990. Gripping description of the terrible hardships endured by the early explorers.

Gage, J. D., and P. A. Tyler. *Deep-Sea Biology: A Natural History of Organisms at the Deep-sea Floor.* Cambridge: Cambridge University Press, 1991. Comprehensive, often technical description of sea floor ecology and denizens.

Gorman, James. *Ocean Enough and Time. Discovering the Waters Around Antarctica.* New York: HarperCollins, 1995. Well-written account of the explorations and science of one of the world's most vital oceans.

Grupper, Jonathan. *Destination Polar Regions.* Washington, D.C.: National Geographic Children's Books, 1999. Great stories about polar explorations.

Helvarg, David. *Blue Frontier.* San Francisco, Calif.: Sierra Club Books, 2006. Strong, general book about the oceans.

Herring, Peter, et al. *Light and Life in the Sea.* Cambridge: Cambridge University Press, 1990. Interesting but sometimes overly technical description of ocean ecosystems.

Hough, Susan. *Earthshaking Science: What We Know (and Don't Know) about Earthquakes.* Princeton, N.J.: Princeton University Press, 2000. Good, up-to-date book about the state of earthquake science, although too technical in places.

Hoyt, Erich. *Creatures of the Deep.* Willowdale, Ontario: Firefly Books, 2001. Fascinating books with strong photography about undersea creatures.

Humphris, Susan E., R. A. Zierenberg, L. S. Mullineaux, and R. E. Thomson. *Seafloor Hydrothermal Systems. Physical, Chemical, Biological, and Geological Interactions. Geophysical Monograph 91.* Washington, D.C.: American Geophysical Union, 1995. Very informative but dauntingly technical summary of vent systems.

Lopez, Barry. *Arctic Dreams.* New York: Bantam Books, 1986. A lyrical, absorbing book about the far north.

Oreskes, Naomi. *Plate Tectonics: An Insiders History of the Modern Theory of the Earth.* New York: Westview Press, 2003. Densely written, informative, sometimes technical discussion of plate tectonics and its evolution as a theory.

Prager, Ellen. *Furious Earth: The Science and Nature of Earthquakes, Volcanoes and Tsunamis.* New York: McGraw-Hill, 1999. Excellent book about the origins and impacts of geologic cataclysms.

———. *The Oceans.* New York: McGraw-Hill, 2000. Good general book with strong pictures, although it deals with some important topics too superficially.

Ritchie, David, and Alexander, Gates. *Encyclopedia of Earthquakes and Volcanoes.* New York: Facts On File, 2006. Loaded with good information, easily accessible and clearly written.

Robison, Bruce, and Judith Conner. *The Deep Sea.* Monterey, Calif.: Monterey Bay Aquarium Press, 1999. Strongly researched, sometimes technical overview.

Scientific American. *Continents Adrift: Readings from Scientific American.* New York: Scientific American, 1970. Jammed with interesting information that offers a glimpse of how crucial scientific theories evolve and develop. The writing is generally dense and sometimes hard to penetrate.

Sullivan, Walter. *Continents in Motion: The New Earth Debate.* Melville, N.Y.: American Institute of Physics, 1991. Good primer on plate tectonics by a veteran science writer.

Sutton, George H., et al., eds. *The Geophysics of the Pacific Ocean Basin and Its Margins.* Washington, D.C.: American Geophysical Union, 1976. A fascinating look at the forces that drive plate tectonics, but definitely written for a hard-core scientific audience.

Svarney, Thomsas E., and Patricia Barnes-Svarney. *The Handy Ocean Answer Book.* Farmington Hills, Mich.: Visible Ink Press, Gale Group, 2000. Interesting, easily accessible collection of surprising ocean facts.

Taylor, Peter Lane. *Science at the Extreme.* New York: McGraw Hill, 2001. This well-written, well-illustrated book does a great job of humanizing science.

Tidballs, Geoff. *Tsunami: The Most Terrifying Disaster.* New York: Carlton Books, 2005. Excellent overview of tsunamis and their impacts.

Van Dover, Cindy. *The Octopus's Garden: Hydrothermal Vents and Other Mysteries of the Deep Sea.* Reading, Mass.: Addison-Wesley, 1966. An absorbing, reader-friendly look at the ecology of undersea vents, now fortunately somewhat out of date.

Wegener, A. *Origin of Continents and Oceans.* 1929. Reprint, New York: Dover, 1967. This reprint of Wegener's classic publication of his initially dismissed and ridiculed theory provides a great look at how scientific ideas emerge and then either advance or fall into disuse.

Wilson, E. O. *The Diversity of Life.* Cambridge, Mass.: Harvard University Press, 1992. A strongly written but technical examination of the importance of a healthy network of diverse species, written by one of the most important scientists who has done work on the subject.

World Spaceflight News. *Undersea Research Photo Gallery and Image Files from the National Oceanic and Atmospheric Administration (NOAA): Sealife, Fish, Coral Reefs, Vents, . . . Submarines, Undersea Habitats, Explorers (CD-ROM).* Progressive Management, 2001. Great images from beneath the sea.

Web Sites

Going Deep: Scientific American Frontiers
http://www.pbs.org/saf/1503/segments/1503-2.htm
The decades of effort that culminated in Alvin *reaching the ocean floor.*

Marine Science, Plate Tectonics
http://www.biosbcc.net/ocean/marinesci/02ocean/mgtectonics.htm.
Colorful source on the distinct layers that make up the planet Earth.

National Oceanic and Atmospheric Administration (NOAA)
http://www.noaa.gov
Federal source of news and information concerning weather, climate, the ocean, research, satellites, the coasts, fisheries, and charting and navigation.

Scripps Institution of Oceanography
http://www.sio.ucsd.edu
Interesting articles under "Scripps News Headlines" and "Explorations" and an informational brochure on preparing for a career in oceanography.

Woods Hole Oceanographic Institution
http://www.whoi.edu
Web site for the nonprofit research facility dedicated to research and education in oceanography. Contains links for online expeditions, news releases, marine science student profiles, and more.

Index

Note: *Italic* page numbers indicate illustrations. C indicates color insert pages.

A

acid rain, from Krakatoa eruption 108
Aegir Ridge 90
Africa
    Arabia split from 120
    in continental drift 6, *10*
    fossils of, South American fossils matching 5
    geological future of 121–123, *124*
    rift systems in 121
    tsunami in 99
Akens, Jim 69–70
Alaska. *See also* Aleutian Islands
    earthquake in *30*, 96
    volcanoes of C-7
Aleutian Islands
    formation of 33, 102
    Pacific trenches and 36
Aleutian Trench 102
Alexander the Great 6
Alps, rock layers of, amount of 11
Althing 85
Alvin (submersible) *62*, *68*
    in East Pacific Rise 67–69
    in Galápagos Rift 26, *50*, *52*, C-1, C-2
    gravity measured by 62
    maneuverability of 67
Amazon rain forest, Andes and 114
Anak Krakatoa 108
Andes Mountains *113*
    characteristics of 110
    in continental drift 111–112
    formation of 102, 112–113, C-7
    Lamb's work in 117
    minerals of 117
    rise of
        climate in 114–115
        observation of 110–111
        trench sediment and 112–113
        warm water influx and 117
    vents of, helium in 112
    volcanoes of 111–112
anglerfish 92
Angus (camera) 50–51, 64

animal migration, magnetic field polarity and 25
Antarctica
    in continental drift *10*
    geological future of 125
Antigua 17
Appalachian Mountains
    rock layers of, amount of 11
    rocks of, Scottish rocks matching 5–6
Aqaba, Gulf of 120
Arabia
    split from Africa 120
    subduction of 123
Arctic ice flow, currents in 76–77
Arctic Ocean *74*, C-6. *See also* Gakkel Ridge
    basins of 73
    characteristics of 73
    in climate regulation 82
    continental shelves in 78
    currents in 76–77, 82
    geological future of 126
    ice cap, thickness of 78
    mineral content of 82
    navigability of 75
    nutrients in 79
    rift valleys of 73
    salt content of 82
    seafloor of
        crust of 81
        geography of 73
        mapping 78
        sample of 79
    seawater composition in 78
    volcanic eruptions in 79, 80–82
Arctic Ridge. *See* Gakkel Ridge
Armstrong, Neil 40
Arnarson, Igolful 85
ash, impact on climate 108
Asia
    Arabia suducted by 123
    geological future of 124–125, *126*
Atacama Desert, nitrogen in 119
Atacama Trench. *See* Peru-Chile Trench
Atlantic Ocean. *See also* Iceland; Mid-Atlantic Ridge
    basin of, Gondwanaland and 6
    currents from Arctic Ocean 82
    depth of 13–14

139

mapping floor of
by HMS *Challenger* 13
during World War II 15–17
Pacific compared to 61–63
seafloor spreading in, rate of 49
seamounts in 16
submersible dives in 41
Atwater, Tanya 20, 28–33
aurora borealis 25, C-6
Australia
in continental drift *10*
geological future of 124
Australian plate, Java Trench and 102
Azores, sea life of 92

B

bacteria
in chemosynthesis 56
of ocean rifts 56–60
symbiotic 57–58
Ballard, Robert 50–51, 54
balloon, in cosmic ray research 38
barrier reef, in Caribbean Sea 17
Barton, Otis 12, 37
basalt, in volcanic pumice 105
bathyscaph 39
Bathysphere 12
Beebe, Charles William 12
Bering Strait, land bridge of 8
bioluminescence, of deep-sea creatures 12
bird migration, magnetic field polarity and 25
black smokers
discovery of 67–70
on East Pacific Rise 67–69
on Gakkel Ridge 81
on Mid-Atlantic Ridge C-4
minerals from 66
salt from 70
in seafloor spreading 70–71
temperatures in 67–69
water circulating through 71
blue whale 6
Brazil, rocks of, South African rocks matching 6
Bullard, Teddy 22

C

caldera 108
California Institute of Technology 29
canyons, ocean. *See* ocean trenches
Caribbean Sea, Puerto Rico Trench in 17
Caroline plate 102
Cascade Mountains, formation of 33
Central Pacific Trough 39
Challenger, HMS 13
Challenger Deep
depth of 35–36, 44
features of 46
human exploration of 36, 42–47

robotic exploration of 47
thermoclines above 43, 46
Chamberlin, Rolling T. 9
chemosynthesis 56
clams, in Galápagos Rift 49, 51–53, 59
climate
Arctic ridges and 82
Krakatoa's impact on 108
in mountain formation 114–115
Coast Guard, U.S. 79
Cocos plate, at Galápagos Ridge 49
cold war, *Trieste* development and 41–42
Columbia River, mud from, in San Juan de Fuca Ridge 26
conning tower, of *Trieste* 43
continent(s)
buoyancy of 2–3, 24–25
growth of 126
subduction of 115
continental drift. *See also* supercontinents
Andes Mountains in 111–112
evidence of *10*, 11–13
force behind 9, 14, 24
in hot spot formation 87
in plate tectonics 31
rejection of theory 9–11, 15
Wegener's theory of 5–8, *7*, 9–13
continental shelves, in Arctic Ocean 78
convection currents
in Earth's mantle 2, 75, 125
hot spots from 87, 90
in magnetic polarity reversal 25
in plate tectonics 75, 125
in trench formation 2
copper, from black smokers 66
coral reefs, of Red Sea *123*, *124*, C-8
core (Earth's) 1–2
convection currents in 75
in hot spot formation 90
magnetic field and 22–23
Corliss, Jack 50, 53
cosmic rays
magnetic field and 25, C-6
ozone layer and 58
Piccard's (Auguste) study of 38
Cousteau, Jacques, in submersible development 39
crater *103*
crust (Earth's) 2
of Arctic seafloor 81
buried, volcanic activity from 91
creation of new 2
hole melted in 89
made by ridges 100
in ridge formation 24–25
in trench formation 2, 125–126
crustal plates
collision of 31–33, 100
development of 2
San Andreas Fault and 31

sediment as lubrication for 113, 115
subducted
　　continents and 115
　　geological formations from 101–102, C-7
　　melting of 101–104
trenches along 125
trenches in middle of 126
currents
　Alvin and 67
　Andes Moutains rise and 117
　from Andes winds 117
　in Arctic Ocean 76–77, 82
　in Earth's core 75
　in Earth's mantle 2
　of oceans 51
Cyana (submersible) 64

D

Dana, James Dwight, on continent and ocean basin formation 8
Dandelion Patch 54
Darwin, Charles 52, 110
Davis, Paul 110, 112, 115
Deccan Traps 87, 88
deep-sea diving
　by Alexander the Great 6
　by Beebe and Barton 12
democracy 85
Devil in the Mountains (Lamb) 112, 117
die-off, during magnetic polarity reversal 25
Dietz, Robert 41
dinosaurs, extinction of 126
Dive to the Edge of Creation (television) 54
diving bell 12
Dominican Republic 17
Dymond, Jack 50

E

Earth
　active zones of 63
　cooling of, radioactivity and 13
　core of 1–2
　　currents in 75
　　in hot spot formation 90
　　magnetic field and 22–23
　crust of 2
　　of Arctic seafloor 81
　　buried, volcanic activity from 91
　　creation of new 2
　　hole melted in 89
　　made by ridges 100
　　in ridge formation 24–25
　　in trench formation 2, 125–126
　geological future of 121–123, 124–125, 126–127
　Griggs model of 14
　magnetic field of. *See* magnetic field
　mantle of 2
　　under Arctic Ocean 81
　　convection currents in 2, 75, 125
　　in hot spot formation 90
　　sample of 75, 79
　spin of, magnetic field and 22–23, 76
　tilt of 75
earthquakes
　in Alaska 30, 96
　in Andes, trench sediment and 112–113
　crust and mantle boundary measured with 102–103
　epicenter pattern of 29
　forecasting 97
　friction in 101
　along Gakkel Ridge 83
　from Krakatoa eruption 105
　from Nazca plate subduction 111
　at Pacific plate, number of 103
　Pacific plate and 39
　in plate tectonics 31
　along San Andreas Fault 27, 27–28
　source of 14
　tsunamis caused by 96–97
　world's worst 108
East Greenland Current 82
East Pacific Rise 61–72. *See also* Galápagos Ridge
　black smokers on 67–69
　expedition to 64, 66–69
　fault lines from 26
　gravity along 62
　hot springs at 53–54
　hydrothermal vents in 64–71
　lava lake in 69
　minerals from 64–66, C-4
　path of 61
　volcanic activity in 64
ecosystems
　in Arctic Ocean 73
　in Gakkel Ridge 81
　in Galápagos Rift 52–59, 57, C-3
　　bacteria in 56–60
　　sulfur in 55–59
　hydrothermal 52–55, 57, C-3
　sulfur-based 55–59
　sunlight in 56
Edmond, Hedy 81
Edmond, John 50
Einstein, Albert 38
Eldfell (volcano), eruption of 93–94
Eldgja Rift, cracking of 85
El Niño 115, 117–119
epicenters, pattern of 29
Erebus, Mount 69, C-5
Eric the Red 84
Eriksson, Leif 85
Eternal Darkness, The (Ballard) 52
Europe
　in continental drift 6, *10*
　fossils from, North American fossils matching 5
　geological future of 123, 124, 126
European Space Agency 25
Everest, George 11

Everest, Mount
    height of 89
    mass of 11–13
evolution
    of humans 121
    of sea life 92
    theory of 52, C-2
Ewing, Maurice 16

F

Farallon plate, collision of 31–33
faults
    from Gakkel Ridge 82
    magnetic stripes and 28–33
    movements of 26–28
    from plate collision 100
    strike-slip faults from 28
fossils
    in Great Rift Valley 120
    matching, from disparate locations 5–6, 31
    from supercontinents 5–6
Foster, Dudley 66–69
Fram (ship) 77
France, submersible development by 40–41

G

Gaherty, James 88
Gakkel Ridge
    black smokers on 81
    crust over 83
    earthquakes along 83
    ecosystems in 81
    fault from 82
    hydrothermal vents in 80
    implications of findings about 82
    plate tectonics tested by 75
    rift valley of 82
    sample from 79
    serpentine in 83
    spreading of 73–75
    volcanic eruption on 79
Galápagos Islands, Darwin in 52, C-2
Galápagos Ridge
    Dandelion Patch site 54
    Garden of Eden site 54–55
    hot springs and 49
    magnetic stripes of 49
    Rose Garden site 55
    tectonic plates at 49
Galápagos Rift 48–60
    chemosynthesis in 56
    clams in 49, 51–53
    ecosystems in 52–59, 57, C-3
        bacteria in 56–60
        sulfur in 55–59
    expedition on 50–53
    hot springs in 52–53
    hydrothermal vents in 52–59
    measuring 49
    sea life in 52–53, C-3

Garden of Eden 54–55
gasoline
    in bathyscaph 39
    in *Trieste* 36, 42, 44, 46
geology, plate tectonics accepted in 33–34
Georgia Institute of Technology 88
Germany, in World War II 15
geysers, of Iceland 92
glaciers, of Iceland 93
global warming, ice cap locking and 78
gold, from black smokers 66
Gondwanaland, theory of 6
Good Friday earthquake 30
gravitometers 62
gravity
    measuring 62
    variations in 62
Great Britain, in World War II 15
Great Rift Valley
    fossils of 120
    Red Sea in 120
    ridge forming in 121
Greenland
    drifting of 90
    exploration of 76
    formation of 87
    Wegener in 11
Griggs, David 14
Guam 42
Gulf of California, formation of 26
Gutenberg, Beno 14
guyots. *See* seamounts

H

Haiti 17
Half Mile Down (Beebe) 12
Harald Fairhair 85
Harvard University 29
Hawaii
    ages of islands 26
    creation of 87
    hot spot of 89
    seafloor of 89, *89*
Hayman, Rachel 66
Healy (icebreaker)
    characteristics of 79–80
    Gakkel Ridge sampling by 79–82
Heezen, Bruce 17
Heimaey, Iceland 93
Hekla, Mount 92
helium, in Andes vents 112
hemoglobin
    in clams 59
    in tube worms 54
Hess, Harry 44
Hillary, Edmund 40
Hitler, Adolf 15, 20
Hospers, Jan 23

hot spot
    force behind 88–91
    formation of 87–88
    in Greenland 87–88
    in Hawaii 89
    in Iceland 87–91
    in India 87
hot springs. *See* hydrothermal vents
humans, evolution of 121
hydrothermal vents
    in East Pacific Rise 64–71
    ecosystems of 52–55, *57*, C-3
        bacteria in 56–60
        in Gakkel Ridge 81
        in Galápagos Rift 52–59, *57*, C-3
        sulfur in 55–59
    factors affecting 81
    in Gakkel Ridge 80
    Galápagos Ridge and 49
    in Galápagos Rift 52–59
    locating 71–72
    origin of life and 58
    in Red Sea 122
    search for 64
    starting and stopping 53–54, 59
    water in 55–56

I
ice lakes, in Iceland 92–93
Iceland 84–94
    democracy in 85
    formation of 87–88, 90–91
    glaciers of 92–93
    history of 84–85
    hot spot of 87–91
    ice lakes of 92–93
    landscape of 92–93
    life in 91–92, 93
    in Mid-Atlantic Ridge 17, 86–87
    mythology of 85, 92
    new islands of 93
    sea life of 92
    settlement of 84–85
    volcanic activity of
        Eldfell eruption 93–94
        Mount Lakagigar eruption 84, 85–87
    volcanic rock of, polarity in 23
Inca 112
India
    in continental drift *10*
    drought in, from El Niño 118
    geological future of 125
    lava eruption in 87
Indian Ocean, tsunami from 95–100. *See also* Java Trench
Indian Ocean Ridge, Red Sea and 120
Indian plate, Java Trench and 102
Indonesia, tsunami in 96–100, 103–104
inner core. *See* core

islands
    formation of
        in Iceland 93
        from Krakatoa eruption 108
        from plate subduction 102
    Hawaiian, ages of 26
    trench formation and 102

J
Japan 36, 124
Java
    in Krakatoa eruption 106–107
    tsunamis in 104, 106–107
Java Trench 95–109. *See also* Krakatoa eruption
    natural disasters from 99–100
    section separated from 96–97
    statistics on 95
jellyfish 54
Juteau, Thierry 66

K
Kaiko (robot) 47
Katmai volcano *103*
Kearney, Andrews 96
Kenya, tsunami in 99
Kilimanjaro, Mount 121
Kin, Eugen 98–99
Knorr (submersible) 50
Koski, Randy 66
Krakatoa eruption 100, *104*, 104–107
    causes of 108–109
    impact on climate 108
    island formed from 108
    volcanoes of 104–105
Kuhnz, Linda 81
Kulu plate, collision of 33

L
Lakagigar, Mount 85–86
Lamb, Simon 110, 112–113, 115, 117
Lamont Geological Observatory 16, 24
land bridges, corresponding fossils and 8
Langmuir, Charles 81
lava
    diverting flow of 93–94
    in East Pacific Rise 64
    in Gakkel Ridge, sample of 80–82
    magnetic polarity in 22
lava flow, in Iceland 86
lava lake
    in East Pacific Rise 69
    in Mount Erebus C-5
League of Nations 77
Leopold III (king of the Belgians) 38–39
life, origin of, hydrothermal vents and 58
limpets 54, 59
Lindemann, Captain 104, 106
Lomonosov Ridge 73

Loudon (ship) 104, 106
Lucy 121
Lulu (submersible) 50

M

Madegascar, tsunami in 99
magma. *See also* core; crust; mantle
    magnetic polarity in 22, 23
    in ridge formation 2, 24–25, 48
    in trench formation 48
magnetic field
    current state of 24, 25
    Earth's spin and 22–23, 76
    ocean ridges and 20
    outer core and 22–23
    polarity reversal of 23–24, 25, 76
    World War II research on 20–22
magnetic polarity
    in lava 22
    in magma 22
    reversal of 23–24, 25, 76
    in volcanic rock 23
magnetic poles 76
magnetic stripes
    discovery of 24
    of East Pacific Rise 63
    fault lines and 28–33
    formation of 100
    of Galápagos Ridge 49
    of Mid-Atlantic Ridge 90
    in plate tectonics 31
magnetite 25
magnetometers 22
Maldives Islands 99
manganese, in Galápagos Rift hot spring 52
mantle (Earth's) 2
    under Arctic Ocean 81
    convection currents in 2, 75, 125
    in hot spot formation 90
    sample of 75, 79
Mariana Islands 36
Mariana Trench 35–47. *See also Challenger Deep*
    submersible exploration of 35–36, 115–116
Massachusetts Institute of Technology 29
Matthews, Drummond 24
Matuyama, Motonari 23
Mauna Kea 89
Maury, Matthew Fontaine 13
Mediterranean Sea
    closing of 122–123
    submersible dives in 40–41
Mendeleyev Ridge 73
Michael, Peter 81–82
Mid-Atlantic Ridge 5–19
    black smoker on C-4
    discovery of 16
    evidence of 13
    geological future of 125
    Iceland in 17, 86–87
    magnetic stripes of 90
    mapping 17–19, *18*
    parameters of 14, 16
    sea life of 92
    submersible exploration of 48
Middle East, oil fields of 123
migration, animal, magnetic field polarity and 25
minerals
    in Arctic Ocean 82
    from black smokers 66, 70–71
    from East Pacific Rise 64–66, C-4
    recycling of 71
    in Red Sea 122
    ridges in formation of 117
mines, magnet-triggered 20
molten rock plumes
    force behind 88–90
    in Iceland 88, 91
Morley, Lawrence 24
Moss Landing Marine Labs 81
mountains
    formation of
        climate in 114–115
        conventional theories on 11
        from plate collision 100
    mass of 11–13
    oceanic. *See* ocean ridges
    oceanic compared to terrestrial xi
    trench formation and 102

N

Nansen, Fridtjof 76–77
National Science Foundation 79
navigation, seafloor mapping in 16
Navy, U.S.
    magnetic field research of 24
    Science Ice Exercise of 77–78
    seafloor exploration by 13–14, 15–17
    Trieste funding from 41–42
Nazca plate
    in Andes formation 112–113
    at Galápagos Ridge 49
    at Peru-Chile Trench 102, 111
    subduction of 111
    tsunami from 111
New Guinea 36
nitrogen, in Atacama Desert 119
Nobel Peace Prize 77
Normark, Bill 62
North America
    in continental drift 6, *10*
    fossils from, European fossils matching 5
    geological future of 125, 126
North American plate
    collision of 31–33
    Java Trench and 102
    Pacific plate and 39, 102
northern lights 25, C-6
North Pole
    exploration of 76
    position of 73, 76

O

ocean(s)
  currents of 51
  depth of 8
  mysteries of 8–9
  positioning in 50
  salt in, origin of 70
  scientific exploration of 13–14
  shadows in 16
ocean basins. *See also* seafloor
  in Arctic Ocean 73
  formation of, conventional theories on 8–9, 11
ocean floor. *See* seafloor
ocean ridges
  crust made by 100
  features of 1
  formation of 1–3, 24–25
    Gakkel Ridge and 82
    heat in 48, 51–53
    in plate tectonics 75, 125
    in Red Sea 121
  global view of 21
  impact on climate 82
  magnetic field changes in 20
  minerals from 66
  spreading 65
  trenches as mirror of 100
ocean trenches
  continental plates stuck in 115
  depths of 21
  features of 1
  formation of 1–3, 100
    heat in 48, 51–53
    in middle of plates 126
    in plate tectonics 75, 125
    weakened crust in 125–126
  global view of 21
  Mid-Atlantic Ridge and 17–19
  as mirrors of ridges 100
  of Pacific, depth of 36
  plate collision at 101–103
  vents in. *See* hydrothermal vents
origin of life, hydrothermal vents and 58
outer core (Earth's) 1–2
  currents in 75
  in hot spot formation 90
  magnetic field and 22–23
oxygen, from photosynthesis 58
oyster bed 54
ozone, cosmic rays blocked by 58

P

Pacific Northwest, seafloor in 31
Pacific Ocean. *See also* East Pacific Rise; Galápagos Rift; Mariana Trench; Peru-Chile Trench; San Juan de Fuca Ridge
  Atlantic compared to 61–63
  basin of 39, 63
  characteristics of 61–63
  seafloor of, exploration of C-1, C-2
  seafloor spreading in 63
    rate of 49
  statistics on 37
  trenches of 36
Pacific plate
  collision of 31–33
  earthquakes at, number of 103
  Java Trench and 102
  North American plate and 39, 102
  South American plate and 111
  subduction of
    earthquakes from 103–104
    trenches from 102
Pangaea 9, 91, 111, 120
Pangaea Ultima 124–125, 126
Parkfield, California 28
Peru, volcano of 116
Peru-Chile Trench 110–119
  Andes Mountains and 102, C-7
  characteristics of 110
  fishing and 114, 115, 117
  nitrates from 119
  nutrients in 115–119
  sediment in, Andes rise and 112–113
Philippine plate 102
Philippines 36
photosynthesis 56
Phuket, Thailand 95
Phyheas of Massalia 84
Piccard, Auguste 37, 38
Piccard, Jacques 35–37
plankton
  in Peru-Chile Trench 115
  sunlight used by 56
plates. *See* crustal plates
plate tectonics
  acceptance of 33–34
  collisions in 100
  convection currents in 75, 125
  discovery of 26–29
  heat in 48–49, 51–53
  in hot spot creation 88–90
  in Iceland formation 91
  impact of theory 33–34
  Mount Saint Helens eruption and 32, *32*
  in seamount formation 44
  validation of 29–33
Plexiglas, for *Trieste* portholes 40
Pompeii worms 71
Puerto Rico 17
Puerto Rico Trench 17
pumice 105
pyroclastic flow 107

Q

quartz, in submersible portholes 40

R

Radcliffe College 29
radioactivity
    in core 101
    in mantle 101
    in rocks, cooling of Earth and 13
Rai Leh Beach 98
Rechnitzer, Andreas 42
Red Sea 120–127, *122*, C-*8*
    civilization and 122
    coral reefs of *123*, *124*, C-*8*
    formation of 120
    hydrothermal vents of 122
    minerals in 122
    as nascent ocean 121
Réunion Island 87
Reykjanes Ridge, Iceland on 86
Reykjavik, Iceland 85
Richards Deep 110
Richter, Charles 14
ridges. *See* ocean ridges
rift valley
    in Africa 120, 121
    in Arctic Ocean 73
    of Gakkel Ridge 82
    hydrothermal vents in 64, 71
    of Mid-Atlantic Ridge 16
rocks
    matching, from disparate locations 5–6, 31
    radioactive, cooling of Earth and 13
Rose Garden 55

S

Saint Helens, Mount 32, *32*
salt, in oceans
    Arctic 82
    origin of 70
San Andreas Fault
    crustal plates and 31
    earthquakes along *27*, 27–28
    East Pacific Rise and 61
    formation of 31
    movement along 26–29
    San Juan de Fuca Ridge and 26–28, 30
San Juan de Fuca Ridge 20–34
    magnetic stripes of 30
    plate collision under 31–33
    Mount Saint Helens eruption and 32, *32*
    San Andreas Fault and 26–28, 30
SCAMP (Seafloor Characterization and Mapping Pods) 78, 79
Science Ice Exercise (SCICEX) 77–78
sciences
    cooperation among 11
    ideal workings of 30
    women in 29
Scotese, Christopher 124–125, 126
Scottish Highlands, rocks of, Appalachian rocks matching 5–6
Scripps Institution of Oceanography
    Atwater at 28–29
    in East Pacific Rise expedition 64, 66
    Galápagos Rift measurements by 49
    magnetic stripes research of 24
    Trieste and 41
    World War II research of 15–16
seafloor. *See also* seafloor spreading
    of Artic Ocean 78–80
    early assumptions about 13
    early exploration of 13–14
    factors affecting 81
    of Hawaii 89, *89*
    magnetic variances in 22
    in Pacific Northwest 31
    topography of, seawater composition and 78
Seafloor Characterization and Mapping Pods (SCAMP) 78, 79
seafloor spreading 24–25, 26–28
    black smokers in 70–71
    of Gakkel-Nansen Ridge 73–75
    in Pacific 63
    rates of 49
    strike-slip faults and 28
sea horses *120*
sea life
    on Atlantic Ocean floor 41
    bioluminescence in 12
    in Caribbean Sea 17
    in Challenger Deep 45–46
    in Galápagos Ridge 54
    in Galápagos Rift 51–55, *57*, C-*3*
    of Iceland 92
    of Mid-Atlantic Ridge 92
    in Red Sea *123*, *124*, C-*8*
sea monsters 6
seamounts
    discovery of 16
    formation of 44
seawater
    composition of
        minerals in 70–71
        seafloor topography and 78
    in deep-sea hot springs 55–56
Sebesi Island 106
sediment, in Peru-Chile Trench 112–113
seismic waves
    crust and mantle boundary measured by 102–103
    molten rock plume measured with 88
serpentine 83
shadows, in ocean 16
Sierra Nevada, formation of 31
silver, from black smokers 66
Smith, Tilly 95–96
Sobolewski, Casey 98
Sobolewski, Julie 98
Somalia, tsunami in 99
sonar, ocean exploration with 16

South Africa, rocks of, Brazilian rocks matching 6
South America
   in continental drift 6, *10*
   fossils of, African fossils matching 5
   tsunami in 111
South American plate 102
   in Andes formation 112–113
   Nazca plate subducted by 111
   tsunami from 111
sphalerite 66
Spitsbergen, fossils found at 5
starfish, in Galápagos Rift C-3
stratosphere, volcanic ash in 108
strike-slip faults, seafloor spreading and 28
subduction
   of Arabia 123
   of continents 115
   of crustal plates
      earthquakes from 103–104, 111
      geological formations from 101–102, C-7
      islands from 102
      melting of 101–104
   in trench formation 100, 125
submarines
   Arctic Ocean explored by 77–78
   magnetic field from 20
   in World War II 15, 20
submersibles. *See also* Alvin; Cyana; Knorr; Lulu; Trieste
   gravity measured by 62
   Piccard's designs for 37–42
   portholes of 40
   testing 40–41
Suess, Eduard 6
sulfur
   in hot-spring ecosystems 55–59
   in Mount Lakagigar eruption 86
Sumatra
   in Krakatoa eruption 106–107
   tsunamis in 97–98, 106–107
sunlight, life sustained by 49, 53, 56
supercontinents. *See also* Gondwanaland; Pangaea
   reassembly of 124–125
   theory of 5–8, *7*
Surtsey 93
Swarm satellites 25
Swiss Institute of Technology 38

T

Tanganyika, Lake 121
Tanzania, tsunami in 99
temperature anomalies, at Galápagos Rift 49, 50–52
Tenzing Norgay 40
Thailand, tsunami in 95–96, 98
Tharp, Marie 17

thermoclines 43–44, 46
Tio (deity) 117
Tonga-Kermadec Trench, seamount of 44
trade winds, El Niño and 118
transform faults 28, 82, 121
transponders, in Galápagos Rift expedition 50
trenches. *See* ocean trenches
Trieste (submersible) C-1
   buoyancy of 36, 42, 44
   in Challenger Deep 35–37, 42–47, 115–116
   conning tower of 43
   development of 40–42
   instruments of 43
   obsolescence of 47
   portholes of 40
   weights used in 40
True North 76
tsunami 99
   deaths from 99
   in Indonesia 97–100, 103–104
   from Krakatoa eruption 106
   in South America 111
   tide during 95
   wave of 97–98
   world's worst 97
tsunami warning system 98
tube worms
   bacteria in 57–58
   in Galápagos Rift 54, 55, 59, *C-4*
   in Gulf of Mexico C-3

U

U-boats, in World War II 15
Ultima Thule 84
United States. *See also* Navy
   Trieste funding from 41–42
   in World War II 15

V

Vatnajökull (glacier) 93
vents. *See* hydrothermal vents
Victoria, Lake 121
Vikings, in Iceland 84–85
Vine, Frederick 24
Virgin Islands 17
volcanic activity
   Arctic Ocean eruptions 79, 80–82
   from buried crust 91
   in East Pacific Rise 64
   factors affecting 81
   of Hawaii 89
   of Iceland
      Eldfell eruption 93–94
      Mount Lakagigar eruption 84, 85–87
   Krakatoa eruption 100, *104*, 104–109
   Pacific plate and 39
   plate tectonics and 32, *32*, C-7
   pyroclastic flow from 107

volcanoes
    of Alaska C-7
    of Andes Mountains 111–112
    of Peru *116*
    trench formation and 102
Von Herzen, Richard 50

W

Wachs, Faye Linda 98–99
Walsh, Don 35–37, 42–47
water
    composition of
        minerals in 70–71
        seafloor topography and 78
    in deep-sea hot springs 55–56
water cycle, saltwater and 70
wave sensors, tsunamis detected by 97
Wegener, Alfred Lothar
    background of 5
    continental drift theory of 5–8, 7, 9–13, 119
    on cooperation among sciences 11
    death of 11

whales
    in Arctic Ocean 79
    bacteria colonies on 59–60
    blue whale 6
white smokers, discovery of 70
Wilson, J. Tuzo 26–28
Woods Hole Oceanographic
    Institution in Galápagos Rift expedition 50, 54
    Red Sea expedition of 122
    Trieste and 41
    World War II research of 15–16
World War I
    Nansen during 77
    Wegener during 9
World War II
    bathyscaph 39
    magnetic field research during 20, 22
    seafloor exploration in 15–17

Y

Yellowstone National Park 87

North Providence Union Free Library
1810 Mineral Spring Avenue
North Providence, RI 02904
(401) 353-5600